Telecommunications Projects Made EASY

by James Harry Green

Published by Telecom Books
An imprint of Miller Freeman, Inc.
12 West 21st St., N.Y., N.Y., 10010
www.telecombooks.com

ISBN 1-57820-009-1

For individual orders, and for information on special discounts for quantity orders, please contact:
Telecom Books
6600 Silacci Way
Gilroy, CA 95020
Tel:800-LIBRARY or 408-848-3854
Fax:408-848-5784
Email:telecom@rushorder.com

Distributed to the book trade in the U.S. and Canada by
Publishers Group West
1700 Fourth St., Berkeley, CA 94710

Manufactured in the United States of America

First Edition, February 1997

Cover design by Robbie Alterio

Table of Contents

Introduction

Eventually, nearly everyone associated with telecommunications becomes involved in project work. The complexity of projects ranges from simple moves and changes of a handful of workstations to replacement of major switching systems or installation of a local area network. Vendors are able to maintain a reasonably high level of proficiency through frequent practice, but their customers may experience a major project only a few times in an entire career. Careers can be damaged by a poorly executed project, so it behooves managers to invest plenty of effort in the planning phases. There is no single or even necessarily best way to plan and execute a project.

This book is intended to help with the tedious parts of project management by listing most of the tasks involved with major telecommunications projects. The tasks listed in this book are not all of the tasks that your project will require, because each project is unique with requirements that are not duplicated in any other project, but this book can help project managers in two ways. First, it serves as a checklist of the elements of common telecommunications projects. Readers who have limited experience in some types of telecommunications projects can use this book as reminder of the tasks that must be performed.

Second, the floppy disk accompanying this book can eliminate much of the drudgery of assembling a project plan. All of the projects discussed are included on the floppy as text files and in Microsoft Project 4.1 form. If you use Microsoft Project 4.1 or later, you can copy the project into your computer, eliminate the tasks that don't apply, add any tasks that are unique to your project, assign them to individuals, and estimate the time. With that, you have a complete project plan ready to publish.

For those readers who do not have Microsoft Project, the task headings are all included in text form. If you're using a spreadsheet or word processor to help manage the process, you can import the text files and save yourself the work of typing them. This method won't give you a complete plan with the tasks linked in sequence, but for simple projects special software isn't necessary.

This book is arranged so that each project plan stands on its own. The first few tasks in each project, which have to do with organization, are nearly the same for all projects, but the rest of the project is unique. The scope of each task is arbitrary, and includeds definitions, explanations, and a few hints. When you prepare your project plan, you may prefer to subdivide some of the tasks or combine them. The main purpose of these explanations is to help you avoid forgetting something important. Anything in a project can be handled if you find out about it in time. The tasks are grouped under logical headings, and are listed in the approximate sequence in which they will be completed. Milestones, which are significant events, are shown in boldface with a milestone icon.

We have used acronyms extensively throughout. These are decoded, but not defined, in Appendix C. For complete definitions, refer to Harry Newton's book ***Newton's Telecom Dictionary.***

This book is intended to grow as the telecommunications industry changes. For example, it does not include a chapter on ATM because we have not yet experienced an ATM project. When the tasks involved in this become clear, they will be added to a future edition of the book. Meanwhile, if any readers care to make suggestions about this book, please e-mail them to harry@pacificnetcom.com. These can be anything from suggestions of tasks to omit or add to project plans, all the way to complete plans that we can include in the next edition.

Chapter 1

Elements of Project Management

The Seven Stages Of a Project:

Confidence

Concern

Panic

Search for the guilty

Punish the innocent

Reward the nonparticipants

For many people, project management is a profession. A project manager doesn't necessarily have to be an expert in all phases of a project to lead the whole thing to a successful completion. A construction project manager may not know how to drive a nail or saw a board provided he or she knows how to marshal the resources of those who do. Many books have been written and courses organized to teach the profession of project management. But what about the person who isn't a professional project manager? Regular managers must bring these resources together when a major change or addition is planned. Careers are enhanced or broken based on how well a project is executed. If the project goes badly, higher management is not impressed by your ability to lay the blame somewhere else. At the conclusion of a disastrous project every one gets splattered as the fan blades whirl.

Ironically, a perfectly executed project won't make you a hero. Users and management expect no less than competence. Even though a major project may come along only a few times during your career, (unless you do project work for a telecommunications vendor), you need to function like a project management expert.

That's where this book comes in.

We'll start by explaining in this first chapter what you need to know about how to plan and organize a telecommunications project. In the rest of the book we'll break projects down into the details. Bear in mind that attention to details makes the difference between success and either failure or, at best, mediocre performance.

Any project consists of the following:

- A series of well-defined tasks...
- that that are performed in a prescribed sequence...
- and assigned to specific individuals...
- with definite completion dates.

Once this structure is set up, the project will succeed, provided all the tasks have been identified and sequenced properly and all the people perform their tasks competently, on schedule, and provide the necessary information to the other others involved in the project. Of course, since Murphy's Law hasn't been repealed, you know that you can't simply launch a project and expect it to go smoothly. You must also establish controls, which are part of any project manager's toolkit. Figure 1-1 shows the steps in organizing and managing a project.

Figure 1-1

Project Modules in an Office Relocation

- Space Remodeling
- Furniture acquistion
- Electrical upgrade
- Telecommunications wiring
- Telephone system moves
- Furniture and office equipment move

The rest of this chapter is devoted to explaining how to do each of these elements.

ESTABLISH PROJECT OBJECTIVES

Telecommunications projects are either isolated projects in which the Telecommunications Department has the leading role, or they are subsidiary projects in which another department is in the lead. An example of an isolated project might be replacement of an existing PBX. Other parts of the company are involved, but they take direction from the Telecommunications Department. A company move to new quarters is an example of a subsidiary project. Moving the LAN is part of the total project, but the Facilities Department is likely in the driver's seat. Whichever type of project is involved, it's helpful to divide the project into modules. Each module is a collection of related tasks that have a definable beginning, end, outcomes, and relations to other modules. **Figure 1-2** shows how an office move project might be divided into modules.

Figure 1-2

Project Modules in an Office Relocation

- Space remodeling
- Furniture acquisition
- Electrical upgrade
- Telecommunications wiring
- Telephone system moves
- Furniture and office equipment move

Each of these modules is subdivided into tasks to form the work breakdown structure, as project managers call it. This book is organized on this basis for the functions relating to telecommunications projects. Within each module objectives must be set and clearly understood by all participants. The key objectives relate to the following:

- Due dates
- Budgets and costs
- Quality of the project
- Major constraints

These variables won't necessarily be part of every project. For example, in the case of a PBX installation, the budgets and costs may have been established by a contract with the vendor, in which case the customer's project manager isn't concerned with costs. To the vendor's project manager, on the other hand, costs will be a critical concern while the customer's chief concern is quality.

Determine Cost Objectives

Anyone can run a successful project if enough money is thrown at it. An effective project manager is conscious of costs and expensive resources, the main of one of which is people. Other resources such as rented test equipment also affect costs and must be managed. Management may have set a budget for the project, and if so, find this out immediately. If not, you may want to set your own budget and keep track of how much is spent.

Determine Due Dates

Every project and module within a project has a due date, and determining due dates is an early order of business. Sometimes the date can be slipped with minimal impact; other times a missed due date means total failure of the project. You must therefore understand the effects of missed dates. If the impact is critical, contingency plans are called for with the amount of planning directly proportionate to the impact of failure.

Develop a Quality Plan

Supposing you are installing a new network operating system over the weekend. Monday morning everything works impeccably except that you failed to establish user accounts for the executive vice president and all of her staff. It was an easy mistake to make: The person downloading the accounts from the old server forgot that one department and no one caught it in the tests. It was a minor omission, but the result was a crisis caused by the failure to plan and make quality checks. Every project must include a quality assurance plan.

Understand Constraints

Besides costs, due dates, and quality, most projects have certain constraints that must be observed. Early or delayed completion may be needed for one department. A group of executives may demand that their telephone numbers not be changed. A crucial decision-maker may be on vacation for two weeks, and unless his approval is received, the project cannot proceed until his return. A project may have to start or finish no earlier or no later than a particular date.

Constraints are unique to every project, but they must be understood while there is time to work around them. Find out what they are, build them into the project plan, and communicate them to everyone.

ORGANIZE THE PROJECT TEAM

As we will discuss in the next section, you must identify all the critical elements in a project, and unless you know these yourself, you need a team of subject matter experts. Many, if not most of the team, will not report to the project manager. They represent the groups that have a working role. Often other groups do not perform work in your module, but belong to the team to coordinate with other groups. **Figure 1-3** shows the composition of a typical project team representing three groups: inside workers, outside workers, and coordinating groups. Representatives of the worker groups may not actually do the work, but they are responsible for its completion.

The roles these people play will be explained as we develop the plans in subsequent chapters. If you aren't aware of all the groups that should be represented in a project, talk to someone who has done it before. The vendor's PBX or LAN installation supervisor is usually a good source.

When the team is assembled, be certain that roles and responsibilities are clearly understood. Establish at the outset who has overall project responsibility. Key vendors will have project managers, but it should be clearly understood that the project management role belongs to the customer. Vendors are there in an advisory capacity and to do the work associated with their part of the project. They know best how to

Figure 1-3

Typical Groups Represented in a PBX Project Team

Inside Workers

- Telecommunications manager
- Information systems manager
- Facilities manager

Outside Workers

- PBX installation manager
- Trainer
- Coder
- Station review coordinator
- Wiring contractor
- Local exchange carrier
- Interchange carrier

Coordinating Groups

- User representatives
- Human resources manager

install and configure their equipment, but they have no direct responsibility to the end users as the project manager does. The project manager may be a contractor. It isn't necessary that he or she be an employee of the customer, but the project manager must have the authority to speak for management. This may mean committing funds and resources of the company to ensure project success. Although this responsibility can theoretically be delegated to a vendor who is doing the physical work, to mix the roles creates a conflict of interest. The project manager must be someone who can make recommendations to higher management with no personal interest except for ensuring project success.

Roles and responsibilities of other team members must be clearly defined. All members must understand that they are responsible for completing their modules within established objectives and for keeping other team members fully informed of status.

DETERMINE TASKS THAT MUST BE ACCOMPLISHED

As we mentioned earlier, a successful project consists of numerous tasks, competently performed, in the correct sequence, and within the objective dates. The success of the project depends on close attention to details. One omitted detail, such as the example we mentioned earlier of failure to initialize a group of key users in a file server, can cause the project to be counted as unsuccessful.

Every task should have a short title that starts with an action word. For example, Install Distributing Frame, Place Telephone Set Designations, and Load Network Operating System are valid task titles.

A key issue you must address is how much detail to track. Take, as an example, a project to install a new structured voice and data wiring system. At the grossest level, you can tell the wiring contractor the start and completion dates and leave all the details to him. If this is a contractor who has consistently demonstrated that he meets his commitments you may not need to track project details any more closely than that.

If a contractor is an unknown quantity, or perhaps a low bidder with a questionable reputation, you may want to track the milestones such as wiring material ordered, material delivered, wire pulling start and complete, wire termination start and complete, testing start and complete, and designation start and complete. If you are the wiring contractor, or a telecommunications manager doing the work with in-house staff, you may track yet another level of detail down to the point of specific rooms being completed, backboards and disturbing frames mounted, ceiling tiles removed and replaced, and other such check points that give you a finer margin of control.

The amount of detail you track boils down to a matter of confidence that you have in the team members and in your own ability to recover if someone fails to deliver. It may also be a matter of how severe is the penalty for failure, and how much slack you have in the project. (Slack is the amount of time available compared to the amount needed to complete the tasks.) It may also be a function of your need to control details. If you are control oriented, bear in mind that unless you are a subject matter expert yourself, no amount of detailed tracking can substitute for competence on the part of the person doing the work.

To determine the tasks needed to complete the project, assemble the subject matter experts and begin listing tasks. If the project is one that some team members have done frequently, a task list will probably already exist. The subsequent chapters of this book and the accompanying disk list tasks that can be cut and pasted into your project plan.

One excellent method of identifying tasks is to start with a row of butcher paper. Cut two or three strips of the paper about 20 or 30 feet long, depending on the scope of the project. Fasten them to a wall to form a large writing surface. Draw vertical lines down the sheet and label these as months or weeks. Draw horizontal lines to form boxes representing the major project modules . Next, the team members begin writing tasks on yellow sticky notes and posting them in the order in which they must be completed. When a module needs an input from a previous module, the group that needs the input puts a different colored sticky note on another group's module in the particular time period. For example, if wiring work cannot begin until the equipment room remodeling is complete, the wiring team would put a blue note in the facilities team's matrix in the appropriate time period. When the team's work is done, the wall is covered with yellow notes representing tasks and blue notes representing linkages to another module. The tasks are then transcribed to a spreadsheet or a project management program such as Microsoft Project.

SEQUENCE THE TASKS

After all tasks have been identified, they are put in the correct sequence. In this part of the exercise, it is important to identify dependencies. The following types of dependencies exist in projects:

- Finish-to-start (FS): The dependent task cannot start until its predecessor has finished. For example, the LAN operating system cannot be installed until the server is operational..

- Start-to-start (SS): The dependent task cannot be started until its predecessor has started. For example, in a PBX project telephone set testing cannot begin until set placement has begun, but testing does not depend on completing the set placement task.

- Finish-to-finish (FF): The dependent task cannot be finished until its predecessor has finished. For example, wire testing cannot finish until wire installation has finished.

- Start-to-finish (SF): The dependent task cannot finish until its predecessor has started.

Not all tasks in a project have dependencies. Some tasks can start and finish independently of others, and do not need to be linked to other tasks. The only constraint to this is that the task cannot start until the project starts, and the project cannot complete until the task is finished.

Dependencies are best displayed using a project management software. Simple projects can be tracked in a spreadsheet, but in complex projects dependencies must be displayed somehow. The two popular methods of displaying dependencies are the PERT chart and the Gantt chart. In a PERT chart, each activity is shown in a box. Arrows connect boxes in the appropriate sequence as shown in **Figure 1-4**. PERT charts can become quite complex in large project projects. Drawing them requires a plotter or they must be drafted by hand. Most project management programs use the Gantt chart as shown in **Figure 1-5**. Activities are linked in a Gantt chart and displayed as shown in the example. While dependencies are not displayed as clearly in a Gantt chart as in a PERT chart, Gantt charts are easier to construct and can be printed on a regular laser printer.

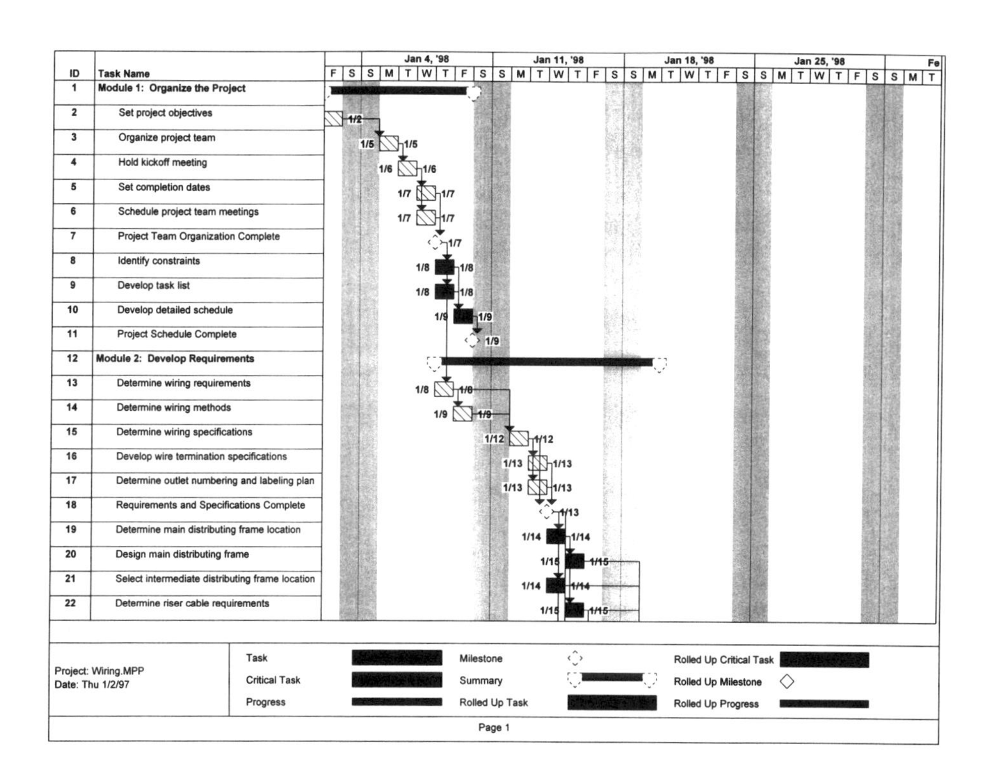

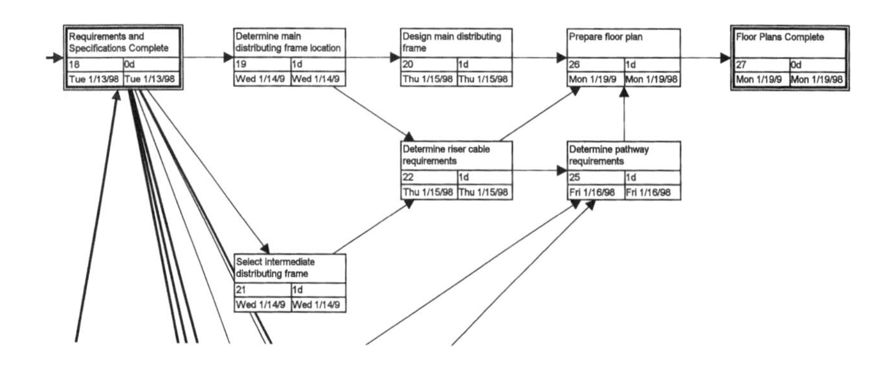

DETERMINE HOW LONG EACH TASK TAKES

Estimating the time for each task is an essential step in developing the overall project time span. Subject matter experts, and those who must perform the work, are asked for their best estimates of the amount of time required. Time estimates are entered into the project management program or written on the PERT or Gantt chart that you used to sequence the tasks.

Sometimes people are reluctant to commit to an exact time schedule, in which case it may be useful to obtain pessimistic, realistic, and optimistic estimates. The pessimistic estimate is the worst case if several things go wrong. The optimistic estimate is the time required if everything falls into place. The realistic estimate is the estimator's best guess of a realistic project time. One technique for combining these is to weight the estimates. Pessimistic and optimistic estimates are multiplied by 1 and the realistic estimate multiplied by 4. For example, if the pessimistic estimate is 12 days, optimistic is four days, and realistic is six days, the weighted estimate would be:

4 x 6= 24

1 x 12= 12

1 x 4= 4

40/5 = 8 days

Project management programs prove their worth in computing elapsed time for the project. When all tasks are sequenced and linked, the program computes the completion date based on the critical path. The critical path is the sequence of linked tasks that represent the longest date span in the project. Project managers must pay close attention to tasks on the critical path because if any of these slip, the completion date slips. Tasks off the critical path can be afforded less attention because they can slip without affecting the project completion date. Bear in mind, however, that slipping a project that is not on the critical path may delay a dependent task that is on the critical path. Project management software allows you to change the time allotted to perform a task to see if it affects the project completion date.

ASSIGN EACH TASK TO ONE RESPONSIBLE INDIVIDUAL

An important point in project management is to assign each task to a responsible individual. The tasks may be performed by a team or a committee, but one individual must be responsible to the project team for reporting status and ensuring that due dates are met. These names are entered in the task sheets in the program, which enables you to produce a complete task list for each person. Be certain that each individual who has assigned tasks understands clearly what is to be done, what due date is required, and what other tasks are precedent to and dependent upon his tasks.

COMPRESS THE SCHEDULE

After all tasks are linked and sequenced, only rarely do they fit in the allotted time frame. Various techniques, such as those discussed below, can be used to compress the schedule.

Review time estimates. Revisit the time estimates of all tasks on the critical path. Ask each responsible team member to consider ways the task can be completed in less time. Look at the optimistic estimates. Ask the task manager what events would have to occur to reach the optimistic estimates. As estimates are revised, pay attention to tasks that were once off the critical path to see if they are now critical.

Work tasks in parallel instead of series. Review the tasks on the critical path to see if some can be worked in parallel with other tasks. If tasks have a finish-to-start dependency, find out if the previous task really must finish before the task can begin, or if some work on the dependent task can start early.

Provide additional resources. Determine whether additional people, test equipment, supplies, or other resources can be provided to compress the time needed for critical path items. Can overtime be worked? Pay close attention to how this affects the budget and quality. The overtime factor is one good reason that vendors performing the tasks should not have overall project management responsibility. The potential for conflict of interest on a fixed-cost contract is too great.

Look for other ways of managing or performing the work. People who have performed a task repeatedly are often blind to the potential of more effective ways of doing it. This is particularly true of work with a large amount of labor content such as installing telecommunications wire. If time permits, purchasing additional equipment, retraining personnel, sometimes a change in the product used can reduce the length of time needed do the work.

Change the due date. When all else fails, the due date may have to be changed. The penalty for missing the due date must be examined in light of other penalties such as spending more than the budget or letting quality slip.

ESTABLISH CONTROLS

Every project must have built-in plans for control. Most project teams are a matrix form of organization in which the participants have a "dotted-line" responsibility to the project manager. Organizationally they report to someone else. Therefore, some of the project manager's controls are indirect. Here is a list of the most common controls and how they are used.

Obtain the commitment of higher management of each organizational unit involved. When diverse groups undertake a project they must be working toward a common goal. If a participant is going to miss a critical commitment, the project manager needs to be assured that the participant's organization is feeling the pressure to perform.

Project meetings. The project meeting with minutes and lists of attendees distributed widely is one of the most effective means of control. When participants make a commitment, it is recorded, and unless the minutes are corrected later, the commitment stands. People who miss project meetings should be identified, particularly when they fail to send a substitute.

As with all meetings, project meetings must have an agenda. All participants are expected to report on critical tasks for which they are responsible. Any apparent slippage in critical tasks must be recorded and the responsible manager required to bring a solution to the next meeting.

The frequency of project meetings is a matter of particular concern. Frequent meetings are required at the beginning of a project while the plan is being created, and toward the end as the due date approaches. In the middle of the project, fewer meetings may be needed, particularly if they are supplemented by reports.

Project plan. A documented project plan is essential to ensure the mutual understanding of all on the team. Project software is an excellent way of documenting the plan, dependencies, and due dates. A narrative plan distributed to all stakeholders is also a valuable tool.

Project reports. Oral project reports are delivered at team meetings. In addition, in intervening intervals, written reports may be required. If all team members have access to e-mail, it is an excellent medium for distributing short written status reports to keep everyone informed. If a report structure is established, omissions should not be tolerated.

MANAGE THE PROJECT TO COMPLETION

With this structure in place, the project is ready for launch. The project manager's chief role is to facilitate information flow. Task leaders are responsible for immediately reporting any deviation from the plan so that adjustments can be made at once. Regular reports and periodic team meetings are needed to keep everyone informed. If the plan is sound and if all participants meet their commitments in performing tasks, the project cannot help but succeed.

At this point we will pause to discuss the use of project management software, which we have briefly mentioned above. Programs such as Microsoft Project are invaluable for controlling large projects. Smaller projects may be managed more easily in a spreadsheet or with pencil and paper.

The chief value of project management software lies in its ability to handle changes with a minimum of manual effort. When the project plan changes, as it inevitably will, the software can recalculate the critical path and completion dates rapidly. Charts and reports can be redrawn and reprinted in minutes with a minimum of manual effort. The program can produce task assignments that clearly state the objectives of the task and its predecessors and dependencies.

Here is how to use this book. All projects are broken into modules. With minor differences, the modules are much the same in all chapters. A flow chart accompanying each chapter shows the modules and how they fit together. In some cases they show linkage to projects in other chapters. The boundaries of the modules are clearly shown in the accompanying text. Some important milestones are also shown. These represent events that mark progress toward completion. Often, milestones must be reached before the project can proceed.

The floppy disk accompanying a book has lists of each task in each project in text form. You can import these into any word processor, spreadsheets, or project management program that can read text files. A more complete project plan is included in Microsoft project files. These files, which have the suffix MPP, show the tasks and milestones. Module titles (which MS Project calls Summary Tasks) are shown in red. Milestones are shown in blue. They are also identified by having a task duration of zero. In addition, tasks are linked in their usual sequence. To convert this into your project, follow these steps:

1. Open the project in Microsoft Project Version 4.1or later. Click on File Project Information, and change the start date to the start of your project. All projects are formatted to start on January 2, 1997. Click on File Properties, and enter your name and the project name in the appropriate spaces. Save the file under a different name to avoid overwriting the template.

2. Delete any tasks that do not pertain to your project. Before you delete a task, look in the Predecessors column to identify any tasks that are linked to it. If successors are linked to a task you are deleting, you will need to break that link and link the successor to another predecessor.

3. Review the sequence of all other tasks. Linkages in your project may be different than the model.

4. Enter the completion intervals. The models are set to the Microsoft Project default of one day. The project team members should be able to estimate the completion times.

5. Identify individuals responsible for task completions and enter these in each task.

At the completion of these five steps you should have a project plan that is virtually complete. You may need to compress the schedule to make it fit, and rearrange some of the tasks, but at this point the preparation is complete and the real work is about to begin.

Chapter 2

PBX Installation and Upgrades

90-90 Rule of Projects

The first 90 percent of the task takes 90 percent of the time, and the last ten percent takes the other 90 percent.

A large PBX conversion is one of the most complex projects a telecommunications manager encounters. Projects involving small PBXs with an uncomplicated trunking structure are not difficult, but the more complex the trunking and the more users that are involved, the more difficult the project becomes. The process is further complicated by the number of vendors involved. The typical PBX project involves not only the PBX vendor, but also representatives from LECs, IXCs, wiring contractors, furniture and office equipment vendors, building trades, and numerous internal organizations that must be included in the planning. When major features such as ACD are added to a PBX, the project becomes more elaborate—to the point that a separate chapter is devoted to ACD planning.

Figure 2-1 shows the major modules into which this project is divided. First, the project team is organized. In preparation for selecting the equipment, the team develops a complete list of requirements and specifications. The team issues a request for proposals and selects the equipment and associated network facilities. When the equipment is selected, the vendor joins the team and participates in detailed planning. The equipment order is placed, and work begins on the equipment room. If new or modified wiring is required, this work proceeds independently of the equipment installation.

The equipment arrival on site is a major milestone, and instigates a flurry of activity proceeding through training, station installation, and cutover. When all the requirements have been satisfied, the customer accepts the systems and completes the project.

This project plan links to other chapters of the book. Voice mail systems are part of nearly every PBX project, and are covered in Chapter 4. Interactive voice response systems are often linked to a PBX, and are covered in Chapter 5. Call accounting is covered in Chapter 8, equipment room preparation in Chapter 12, and wiring systems in Chapters 13 and 14. The equipment room and wiring system modules are shown in dotted lines for clarity, but they are not part of this project. Finally, Appendixes A and B cover system acceptance and security.

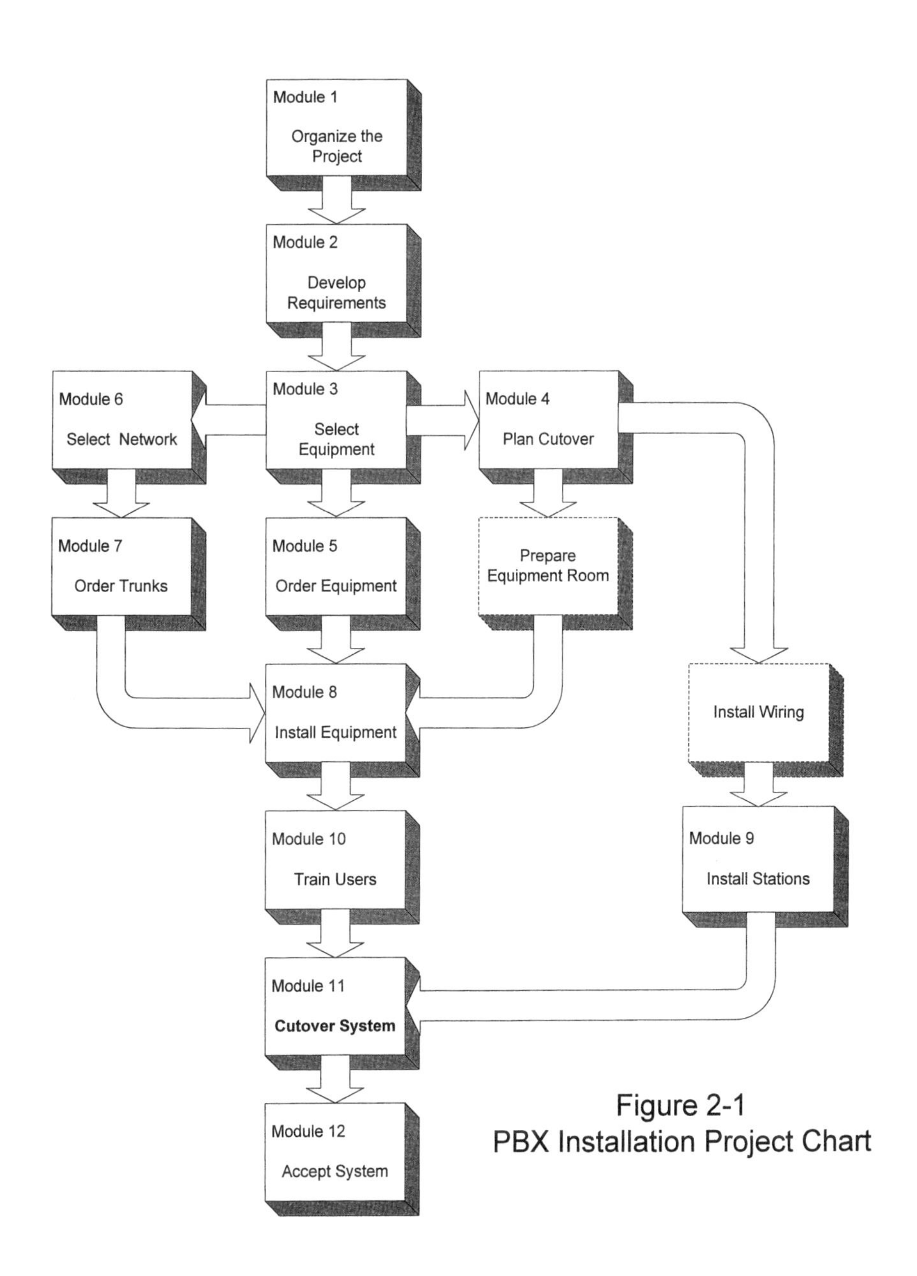

Figure 2-1
PBX Installation Project Chart

Module 1: Organize The Project

Set project objectives

Determine key objectives for the project. These must include completion date plus dates for any intermediate tasks that may affect target completion dates. Determine budgets for the project.

Organize project team

Identify all who are needed in the project. Representatives should be identified from the PBX vendor, which should have a project manager who can commit the company's resources. The LEC and the IXC should have representatives, although they may not be required at all meetings. If rewiring is involved, the wiring contractor should provide a representative. Identify internal departments that must be represented. Typically, these include the telecommunications, information systems, facilities, and in some projects, human resources and public relations. Be certain that roles and responsibilities of all team members are clearly understood and accepted.

Hold kickoff meeting

In the initial project meeting all team members should understand their own and others' responsibilities. The kickoff meeting has the following objectives: assign and accept key responsibilities, communicate objectives and constraints of the project, establish schedules and content of reports and project meetings. Someone should be assigned responsibility for preparing minutes and distributing them within a day or two of the meeting.

Set cutover dates

Determine key dates in the project, including the final cutover dates as well as any interim dates that must be met.

Schedule project team meetings

Develop a schedule and place for future team meetings. The frequency of meetings depends on the competence of the project team, the complexity of the project, the penalties for missing due dates, and the numbers of people who must be kept informed.

Project Team Organization Complete

Identify constraints

Determine from all project team members any constraints that their company has with respect to force availability, inability to work certain dates, availability of key personnel or test equipment or other factors that may affect project completion. Document all such constraints in the minutes

Develop task list

Develop a detailed list of tasks that must be performed. Assign each task to a responsible individual. Sequence tasks in the order in which they must be performed. Obtain estimates of the amount of time required for each task.

Develop project schedule

Identify milestones in the schedule. Set dates for completion of each milestone. Determine the critical path and determine whether the schedule fits within the objective interval. If it does not, determine how to compress the schedule until it fits. Create Gantt and/or PERT charts showing task sequence and schedules, and distribute to all team members.

Project Schedule Complete

Module 2: Develop Requirements

Develop station port requirements

Determine the number of station ports of different types (analog, digital, BRI) that are required at cutover.

Determine growth projections

Determine how large the system is expected to grow during its life. Establish requirements for ultimate capacity in both station and trunk ports.

Determine equipment room requirements

Determine dates by which the equipment room must be ready. Outline the work that must be done to prepare it. For extensive equipment room work, refer to Chapter 12

Develop station numbering plan

In a single-PBX environment the station numbering plan will be driven by the availability of DID numbers. In a multi-switch environment the numbering plan will be more complex. Determine how many digits will be dialed. In a multi-switch environment a leading digit is often required to direct the call to the appropriate switch. Be certain the PBXs are equipped with uniform dial plan software.

Determine protection requirements

Station protection is usually required for inter-building cable. Station protectors installed across cable pairs protect personnel from electric shock and equipment from damage in case of cross with power or lighting strikes. If existing cables are not protected, check with a protection authority to determine if it should be added.

Determine overhead paging requirements

Determine whether overhead paging is required. If so, determine the number of zones. Arrange for a contractor to design speaker locations in accordance with requirements.

Determine PBX networking requirements

In a multi-PBX environment, networking software may be required to provide feature transparency across the network and to enable multiple locations to share the same voice mail. If switches of different manufacture are involved, networking will not be possible unless the systems support the Q-Sig protocol. Check to be certain that the appropriate networking software is included in the PBX order.

Determine power fail transfer requirements

Determine whether power fail transfer has been proposed, and if not, whether it is needed. Note that PFT does not work with digital trunks. Determine how many trunks should be equipped, where emergency phones should be located and how they will be wired.

Establish system security policies

Several policies are needed for the vendor to set up the PBX. These include whether dial-up ports will be protected with external devices such as dial-back modems, frequency of forcing password changes on PBX ports and voice mailboxes, minimum password length, and whether certain features such as remote access and off-system forwarding will be allowed.

Determine voice mail requirements

Determine where automated attendant functions will be used, how many menus, and how long they will be. Determine default number of rings to cover to voice mail. Determine whether different coverage paths will be used for internal versus external coverage. Determine whether and how voice mail will be backed up. Identify any networking requirements.

Determine music-on-hold requirements

Most PBXs play music or announcements while callers are on hold. Determine what will be played, source of music, and whether the same material will be heard by all stations.

Determine attendant console requirements

Determine feature requirements of the attendant console such as busy lamp field, direct station selection field, point-and-click call transfer, forward to auto attendant on busy, etc. Verify whether multiple consoles or overflow to another station are required. Determine whether centralized attendant service is required.

Determine ACD requirements

Determine whether ACD is required. Determine number of groups or splits, how many agents, what type of telephone set will be used, how many supervisor's terminals will be required and where they are to be mounted. Determine announcement requirements. Determine call flow. See Chapter 3 for ACD planning information.

Determine fax requirements

Determine requirements for personal fax lines such as those built into modems, and for common area fax lines. Determine if these will be assigned to analog ports. If so, ensure that the fax number is either compatible with the station numbering plan, that it can be translated to a compatible number within the PBX or that the number can be changed.

Determine modem requirements

Identify all modems that will be assigned to PBX ports. Determine if the modem must both send and receive calls. If receiving is required, assign a DID number. Check to ensure that the modem is suitably isolated from the network to prevent hackers from reaching the LAN.

Develop training requirements

Determine what courses will be presented, how many people per class, the number of classes that are needed, what types of telephone sets will be used for training, and how classes will be scheduled and participation encouraged. Training should include instructions on sets and features and on the attendant console. Training may be included for PBX and voice mail system administration and for special features such as ACD and IVR. See Chapter 3 for ACD training information and Chapter 5 for IVR training information. Training may also be required on call accounting system administration (see Chapter 8).

Determine earthquake bracing requirements

If the PBX is being installed in an area that is susceptible to earthquakes, bracing is highly recommended. Even in light earthquake areas, bracing prevents cabinets from shifting position or tipping over in case of earth tremors.

Develop ARS plan

Determine which trunk group handles various types of calls, and how calls overflow from one group to another when the group is busy or inoperative. Determine requirements for digit insertion and deletion, which may be required for FEX trunks or when completing long distance calls over a tie line group to a distant switch.

Develop synchronization plan

All devices on a digital network must be synchronized. Devices such as PBXs and routers take their timing from the source closest to the national reference frequency. Other devices on the network slave from the master. Be careful to avoid synchronization loops. Prepare a synchronization plan. Consult an expert if in doubt about plan validity.

Identify effects on customer directory

If stations are added, fax and modems moved from business lines to analog ports on the PBX, extension numbers changed, DID numbers added, or any LEC or 800/888 numbers changed, the customer directory will be affected. Determine what must be done to reprint and distribute the directory.

Identify effects on stationery

If any station numbers are changed, DID added, e-mail addresses changed, or other changes that affect business cards, letterheads, and other printed material such as brochures, users must be informed. Assign new numbers as early in the process as possible so users can get their stationery reprinted.

Develop intercept plans

When telephone numbers are changed, provisions are often required for intercepting calls to the old number and redirecting them to the new. The LEC will usually transfer calls if the old number was listed in their directory and the new number is furnished by them. DID number changes are normally the responsibility of the customer. The LEC does not intercept or refer calls on disconnected DID numbers. These can be handled by forwarding the old number to voice mail if one exists. Although no transfer of calls is possible, disconnected and nonworking numbers can be forwarded to a digital announcer that states that a nonworking number at XYZ company has been reached.

Develop internal publicity plans

Determine what kind of publicity should be disseminated through the organization to prepare people for the cutover. Publicize the key dates. Consider the need for departments to participate in station reviews, effects on all printed material, possible disruption to service during the transition, any need for number changes, need to identify critical stations, freeze dates and their purpose, etc. Determine how information will be disseminated; for example, e-mail, bulletin board, employee newspaper, etc.

Requirements and Specifications Complete

Module 3: Select Equipment

Prepare RFP

Write a request for proposals or quotations listing all requirements and desirable features of the system. Issue the RFP to selected vendors. Set a due date for responses.

Develop evaluation criteria

Determine how the product will be selected. Price is always a factor, but other factors should be considered such as compatibility with existing equipment, product support capabilities, ease of use, ease of management, and references from users with similar configurations.

RFP Responses Received

Evaluate responses

Vendor responses are reviewed. Non-conforming products are eliminated. Rank the remaining proposals and select the top two or three products for further analysis.

Demonstrate products

The vendor demonstrates the product in an environment similar to the one in which the products will be used. Watch demonstrations of major features. Evaluate the products against the evaluation criteria.

Check references

Discuss with at least three customers their experience with the product. Determine whether the vendor has been sufficiently responsive in problem cases. Get their opinion on ease-of-use in day-to-day operation. Ask whether they would buy the same product again.

Select the product

Choose the winning proposal. Obtain management concurrence. Arrange financing, if appropriate. Negotiate a contract with the vendor.

Equipment Selected

Module 4: Plan Cutover

Develop change control procedures

Lack of change control is one of the most common reasons for failure of a cutover. The switch database must be programmed in advance of the cutover. If possible, the vendor programs the new switch to the maximum degree with information dumped from the old database. If change control procedures are not exercised, the risk is high that the new PBX will not reflect the station configuration as it existed in the old one. The most effective way to control changes is to establish a freeze date beyond which new orders will not be processed. If changes must be made in the old switch after the freeze date, a careful record must be kept outside the new switch so the same change can be worked twice: once in the old and once in the new switch. Announce the freeze date well in advance so users have a chance to get their changes made before programming of the new PBX begins. Accumulate requests after the freeze date, and work them after the cutover is complete. If trunks are being reused, control of trunk changes is at least, if not more critical than station changes. Trunks can be added or deleted provided the information is processed in the new switch after the cutover. With changes of all types, it is imperative that the new PBX keep pace with any changes in the old one.

Develop cutover methods

The cutover plan relates to the method of making the physical transition from the old system to the new. Numerous issues must be resolved in the cutover plan. Determine whether the cutover will be phased or "flash." Will stations and trunks be reused? Is any existing equipment being reused? Determine how turnaround space will be obtained in the equipment room. Determine whether the cutover will be a weekend or a weekday cut.

Develop station wiring plan

Determine whether new or reused wiring is involved. If new wiring is required, refer to Chapter 13 for details on developing a wiring plan. If wiring will be reused, determine whether jacks are wired and designated properly.

Determine telephone set installation methods

In large cutovers involving numerous telephone sets, placing all telephone sets and changing all cross connections in a single weekend may be a difficult task, particularly if service continuity of a significant number of stations is required. If spare cable pairs to the work stations are available, it may be feasible to install new sets ahead of time. This leaves users with two telephones for an interval, but spreads out the job of replacing sets.

Develop ACD cutover plan

Develop call flow and routing into the ACD. See detailed ACD planning in Chapter 3.

Develop voice mail cutover plan

Develop a plan for getting voice mail users trained to use the system effectively. Discuss default greetings and prompts. Determine station number assignment for the voice mail remote access number. Discuss security methods. See detailed voice mail planning in Chapter 4.

Determine trunk access codes

Single-digit codes such as 8 and 9 are used for trunk group access through the automatic route selection. Trunk access codes of two or more digits are used for direct trunk group access, bypassing the automatic route selection. Trunk access codes are chosen, and, if possible, restricted to a narrowly-defined class of service. Make certain these codes are not accessible through voice mail because of the hazard of toll fraud.

Develop QA plan	Develop a plan for quality assurance. Include such factors as checking that all stations are included, checking reused wiring for bridged tap, verifying station restrictions, verifying ARS accuracy, verifying compliance with manufacturer's specifications, etc.
Determine back door access method	Every voice mail system needs an extension number for users to check for messages. A DID number may be assigned for checking from off-site without attendant assistance. Some systems may require an 800/888 number for remote access.
Develop trunk testing plans	Determine how trunks will be tested. Determine what tests will be made, who will make them, and what records will be kept of the results.
Develop station testing plans	Determine how station operation and feature operation are to be handled. The vendor usually sets up a testing center equipped with display telephones. As testers complete the prescribed list of tests at the telephone, they call the center, which answers the phone with the telephone number of the calling station. The tester verifies the number against the designation on the set, and hangs up. The center calls back the station to verify that the ringer works. The center checks off each station on a master list to ensure that all stations have been verified.
Develop trouble reporting procedures	For the first few days following cutover, a help desk will be established and employees will be informed on procedures for reporting trouble. Procedures should also

include methods for referring troubles to technicians, maintaining a trouble log, and clearing trouble with the users when clearance is reported to the help desk.

Identify critical stations

Most cutovers occur over a weekend when stations are not in use. Many organizations such as hospitals, hotels, and companies with 24-hour-per-day shift operation have stations that must remain in service. If new trunks are being installed, the change must be coordinated with the LEC for incoming calls. If trunks are being reused, incoming calls will route to the new switch as soon as trunks are transferred. Outgoing calls can usually be accommodated by splitting the trunk group between the old switch and the new to provide outgoing service for whichever PBX the station is connected to. Critical stations must be monitored for conversation and transferred as quickly as possible. The interruption should normally be only a few seconds. Coordinate the method of handling critical stations with the users to make sure that everyone understands the process.

Prepare floor plans

Floor plans of all assigned stations are needed. A floor plan of the equipment room and a backboard plan are also needed. Determine who will prepare the plans, what kind of detail they will include, and when they must be done. Remember to include plans for common area telephones. Station user locations must also show personal fax and modems on the plan.

Floor Plans Complete

Develop station restriction plan

Station classes must be assigned. Generally, stations installed in public areas such as waiting rooms and lunchrooms are restricted to local calls. Executive stations may be completely unrestricted. In between, different classes of service may be used to limit some stations from international calls, and some may be restricted to placing tie line calls, but no toll calls. Also determine what features will be assigned to various classes. For example, some stations may be permitted to use off-system forwarding, which is vulnerable to toll fraud.

Compile station records

Develop a complete record of existing stations. The record should include extension number, user name, room or cubicle number, department, and line location on the existing PBX.

Assign station numbers

Develop an old number-new number list showing all existing stations plus any that are being added with the project. Show the existing and new line port numbers. Provide the list to users in time to enable them to reprint stationery and publicize any new DID numbers.

Station Planning Complete

Set station change freeze date

To control the project effectively, a date must be set, beyond which no further station changes will be accepted in the existing switch if the new switch is to obtain database information from the existing switch. If the cutover is to a new PBX, the database must be frozen while the new one is created. Publicize the need for the freeze to all station users and inform network them of the last date for placing orders.

Set change freeze date

Develop a date beyond which no further trunk changes will be ordered. The network freeze date applies to local, IXC, and tie trunks, as well as to any voice mail trunking.

Design system software

The vendor's representative develops coding instructions for all variables in the system software, for both stations and trunks.

Develop contingency plans

Begin contingency planning by determining the effect of missed due dates. Next, think of all that can go wrong, and determine how it would be dealt with. If the due date can be slipped without penalty, no contingency plan may be needed. If a missed date causes a severe penalty, consider other alternatives such as overtime, using a temporary service, escalating to higher authorities, etc. At a minimum, develop contingency plans for missed trunk due dates, failure of the switch or a critical component to arrive on time, "acts of God" such as fire, flood, civil disturbance, etc. Evaluate the likelihood of events occurring, and plan accordingly. Also consider whether an opportunity to advance the schedule may arise.

Cutover Planning Complete

Module 5: Order Equipment

Conduct station reviews

During station reviews the vendor's representative meets with each user or departmental representatives to determine what type of station set will be provided, station coverage paths, class of service that will be assigned, features assigned, etc. Use the output of the station review to develop the final station configuration.

Develop final voice mail configuration

Determine final number of ports, port configuration, and hours of storage for voice mail.

Develop final PBX configuration

Determine final number and types of sets. Determine final feature list. Determine final quantity and type of station and trunk ports.

Prepare equipment order

The PBX equipment order is assembled, using information and prices from the vendor's proposal and modified by changes made as a result of the final design and the station review. Consider that the vendor's proposal is usually a response to a pro forma design used to compare prices among vendors, and does not represent a final design. Obtain final approval on the equipment order.

Order equipment

Place the equipment order with the manufacturer. Obtain commitment from the manufacturer on delivery dates.

Module 6: Select Trunking

Determine local trunking requirements

The quantity and type of trunks must be determined. Types include primary rate ISDN, two-way and DID CO trunks, which may be either analog or digital. Determine signaling methods on trunks (usually ground-start and DTMF on local CO trunks). For digital trunks determine whether conventional D-4 or ESF/B8ZS framing will be required. ESF/B8SZ is always required for PRI trunks. To determine trunk quantities, review traffic on the existing switch, if one exists. If no existing information is available, someone with experience in trunking design should be able to estimate the quantity required. Determine whether some trunks will be one-way. Determine hunting patterns.

Determine PRI requirements

Determine quantity of primary rate trunks and features they will be equipped with. Features include call-by-call service selection, calling party identification, etc.

Determine digital trunk requirements

Determine quantity of digital trunks. Determine how T-1 groups will be segregated as one-way, two-way, DID, etc.

Determine analog trunk requirements

Determine quantity of analog trunks needed. Determine how trunk groups will be segregated as one-way, two-way, DID, etc. Determine which trunks will be assigned as PFT.

Determine interexchange carrier trunking requirements

IXC trunks are carried on T-1 with rare exceptions. Determine the quantity of trunks based on expected long distance and 800/888 usage. Determine signaling methods on trunks (usually DTMF wink start unless PRI is used). Determine whether the T-1 will be segregated for incoming and outgoing usage or if trunks will be two-way. If video conferencing is involved, PRI may be required. Determine degree of IXC involvement in the LEC portion of the trunks: furnish total service, coordinate only, or customer furnishes LEC portion. See Chapter 10 for additional information on WAN projects.

Determine 800/888 service requirements

If the company has existing 800/888 numbers, these must be transferred to the new PBX. If they are business line 800/888 services and T-1 access is being installed, the IXC transfers the numbers to the new trunk group and provides DNIS numbers. Obtain a list of all existing 800/888 numbers. Determine need for DNIS numbers. Determine need for additional 800/ 888 numbers .

Determine tie line requirements

If the PBX is being installed in a multi-switch environment, tie lines to the other PBXs may be required. Determine whether tie lines will be required, and if so, what quantity and type they will be (analog, T-1, etc.). Determine whether tie lines will be shared with data. Determine signaling methods (usually E&M/DTMF).

Determine trunk transition methods

When an existing PBX is being replaced, existing trunks can be transferred to the new PBX, or all new trunking can be ordered. If the project involves a change from analog to digital trunking, the new digital trunks can be ordered so they are available for testing at least a week ahead of cutover. If existing trunks are being transferred, trunk testing is not required if all trunks are known to be operational.

Determine DID requirements

When DID is being installed for the first time you must obtain a block of numbers from the LEC. More numbers than you need immediately are usually obtained for future growth. Determine whether numbers out of the DID sequence are needed to transfer analog devices such as fax and modems to analog ports on the PBX without number change. Since it is desirable that the DID number be a function of extension numbers, a change in the extension numbering plan may be required.

Trunk Planning Complete

Module 7: Order Trunks

Order local trunks

Determine how much lead time the LEC requires from order to installation date. Place orders with plenty of time for the LEC to ensure that facilities are available. Check to ensure that the entrance cable has sufficient capacity. Be prepared to furnish requirements to the LEC for signaling type, trunk quantities and type, and required due dates. Set the due date at least one week prior to cutover to provide time for testing.

Order tie lines

If tie lines are required, place orders with the LEC or the IXC that will furnish them. Be prepared to furnish trunk quantities and type, signaling type, and due dates. Set the due date at least one week prior to cutover to provide time for testing.

Order IXC trunks

Place orders for the IXC trunks, which are usually T-1. Specify type of trunk, signaling type, quantity of trunks, number of channels reserved for incoming service (if two-way service is not used), and due date. Set the due date at least one week prior to cutover to provide time for testing.

Notify IXC of all new trunks

Inform the IXC of all trunks that are being added with this project so they can be included in the IXC's billing records.

Order 800/888 services

Order 800/888 services from the IXC. State the quantity of numbers and DNIS requirements. For business line service, indicate the telephone numbers to which they terminate. Order 800/888 numbers transferred from existing to new trunks, specifying to the IXC the time and date on which the transfer must be made.

Trunk Orders Complete

Module 8: Install Equipment

Prepare equipment room

Get the equipment room ready for equipment installation. See details in Chapter 12 .

Equipment Room Complete

Deliver PBX

The vendor has the PBX delivered from the factory. The length of time for this item is the manufacturer's interval from receipt of order to delivery.

Equipment On Site

Unpack and position cabinets

The PBX installer unpacks the cabinets, inventories the system to be sure all parts have arrived, and puts the cabinets in position. If earthquake bracing is required, it is installed and cabinets are bolted to the floor.

Connect power plant

For AC powered systems the PBX is connected to the UPS supply or directly to AC power. For battery operated systems the rectifier is connected to AC power. Batteries are installed and the rectifier is adjusted to bring the batteries up to full charge. For wet cell systems the cell voltage and specific gravity are recorded after the battery supply is fully charged.

Connect system to power

The system is connected to AC or DC power and to ground. Be certain that the ground meets all manufacturer specifications for tightness of connections, wire gauge, and bonding to power.

Install cable supports

Install cable trays or racking to support cable between the cabinets and between the cabinets and the MDF. Cables must be lashed to the cable rack and separated physically according to the manufacturer's specifications.

Power up system	Power is applied to the system and all of the manufacturer's installation tests are run. The installer checks completion of all items on a master check list .
Load system program tape	The generic program is loaded. System initialization tests are performed and recorded on a master check list. All system diagnostics are run and a record of their completion is maintained.
Test battery backup	AC power is disconnected to the battery or UPS supply and left disconnected long enough to verify the capacity of the system. A record is made of the elapsed time from disconnect until the system is incapable of call processing because of low voltage. For DC systems the bus bar voltage at the system failure point is recorded.
Input system databases	Vendor installs trunk and station database. A complete database record is printed and maintained.
Program ARS	The ARS is programmed according to instructions provided in the ARS plan. All programming must be tested to ensure that calls are directed to the right trunk group, that they overflow to secondary groups when all trunks are busy, and that the correct digits are inserted and deleted.
Install and test trunking	Where new trunking is being added, the LEC and IXC install trunks. Analog CO trunks are tested for dial tone and signaling. Install CSUs on T-1 Trunks. LECs and IXCs loop back trunks and test for integrity and error-free performance.

Trunks Installed

Connect trunks to system

The PBX installer connects trunks and verifies operation and hunt sequence. Check all analog DID trunks individually with the central office.

Test network features

Test all features that are included in network orders. These include call-through tests of all trunks, test of features such as call transfer, call-by-call service selection, trunk hunting, forward on busy or no answer, etc.

Install attendant consoles

Connect the attendant console to the switch and test for operations of all features. In a multi-console environment, test to be sure that calls are distributed between consoles according to the specifications. In centralized attendant service, verify that incoming calls at remote locations forward properly to the CAS location. Verify that the console attendant sees the proper trunk group designations on the alphanumeric readout.

Install voice mail

Turn the power on for voice mail and observe that the system boot-up occurs properly.

Perform installation tests

Perform all of the initialization tests required in the manufacturer's manual. Prepare a written check list to verify that all tests were performed and provide the list to the customer.

Installation Tests Complete

Program default greetings

The vendor or the customer must record greetings that the voice mail system requires. This may include dial-by-name instructions, and any greetings that the manufacturer permits the customer to change.

Program voice mailboxes

Voice mail boxes are programmed according to instructions. Use caution in programming any voice mail boxes associated with inactive stations because of the risk of capture by hackers.

Program station restrictions

The classes of service are programmed in accordance with the plan. Restrict the use of certain features that are toll fraud prone. Access to hacker-vulnerable country codes, area codes, 800 numbers, and 900/976 numbers should be restricted in accordance with the plan. Restrictions are tested to ensure that programming is correct.

Restrict remote access

Unless the organization has a clear need for remote access, it should be disabled. If it is enabled, check to be sure that password length is consistent with industry security practice.

Block transfer from voice mail to outgoing trunks

Most toll fraud results from callers transferred through voice mail to an outside trunk. The manufacturer's method of preventing this must be installed. Make certain that the only transfer allowed through voice mail is to a valid extension number, and that transfers to trunk access codes, including codes to tie lines to other

PBXs, are denied. Test denial of all trunk access codes.

Program administration terminal security

Determine the administration terminal password. The password should be long enough to make it difficult to hack. Security is improved by having one or more special characters in each password. If a remote terminal security device is being installed, program and connect it. The terminal is connected to a business line. DID ports should be avoided so the terminal can be accessed if the switch is down. If a separate voice mail port is involved, the same provisions apply except that the voice mail terminal can be assigned to a DID number.

Connect call accounting system

If a call accounting system is used, connect it to the SMDR port and program it. See Chapter 8 for details in setting up a call accounting system.

Test power-fail transfer

Disconnect power to the system and verify that power-fail transfer circuits work.

Perform final switch tests

At this point the switch is installed and ready to cut over. Final system inspections should be performed as described in Appendix A. If time permits, the PBX should be allowed to burn in without traffic for at least a week to screen out any defective components. Any final tests recommended by the manufacturer's installation manual should be performed at this point.

Module 9: Install Stations

Install wiring

If new wiring is required, or if existing wiring must be modified, complete wiring before station set installation begins. See Chapters 13 and 14 for wiring system planning.

Prepare station installation floor plans

Mark a set of floor plans for the installation crew to use in placing telephone sets. The floor plans must show extension numbers of the primary station plus any additional ports that are provided for faxes, modems, BRI, etc. If extension numbers are being changed the floor plan should be marked with both the old and new numbers, or an old-number-new-number list is needed.

Provide staging area

In projects that extend over several days, the vendor will require a lockable space for storing telephone sets and some installation materials. Identify space in the building large enough to contain the equipment and materials. Note that equipment room storage is generally undesirable because of the flammable the and dust-generating nature of packing material.

Designate station sets

When new sets are being placed, boxes are opened and designation strips put on sets. The room or cubicle number is written on the outside of the box. Sets are grouped by area for delivery at cutover time.

Place station equipment

After cutover to the new system, the new station equipment can be activated. If sets are in place, they can be tested immediately, but if sets are placed during the cutover period, the set placement process must be coordinated with the process of running crossconnects. Leave users' manuals at each station unless they are being provided during the training classes.

Call-through test each station

After stations are installed, technicians call a central dispatch center which reads the extension number back to the technician, who verifies that it matches the designation on the set. The technician hangs up and the center calls the station back to verify ringing. A master check list is kept of all stations tested and all trouble cases. At the vendor's option, troubles are fixed on the spot or logged for repair by a different work force.

Test station coverage path

Each station's coverage path is tested by leaving the station off-hook and checking to be sure that the call goes to voice mail, the attendant console, or other designated coverage location.

Set up help desk

A help disk staffed with trained personnel is set up to operate for the first few days after cutover. The purpose of the help desk is to accept trouble reports from users, clear them if possible by providing telephone assistance, and dispatch trouble when it cannot be cleared over the telephone. The help desk keeps logs of all reports received and cleared. When the quantity of reports drops to an acceptable level, the help desk reverts to normal company operations.

Station Installation Complete

Remove old telephone sets

All telephone sets replaced during the cutover are removed, boxed, and stored in a location of the customer's choice.

Module 10: Train Station Users

Determine training room location

Space must be provided and scheduled for the training room. In large organizations multiple classes may be held simultaneously, requiring multiple rooms.

Develop training schedule

Training classes are scheduled and participants notified. Consider what method will be used to ensure that everyone attends.

Set up training room

The PBX vendor wires temporary station cables into the training room(s), and connects telephone sets. An appropriate number of sets of each type that will be used are installed so users can be trained on the type they have on their desks. Special provisions may be necessary for ACD users and users of special applications such as data switching. Set up the attendant console in a convenient location for training the attendants. If on-site administrative training is being provided by the vendor, provide space for this.

Train station users The vendor's trainer delivers classroom training as agreed to in the training plan. Pass out user manuals if they are being provided during training as opposed to being left at the station at cutover. Attendance records are maintained to have a record of those who to fail to attend. Inform people of how to reach the help desk.

Train console attendants Train console attendants on how to answer, transfer, and queue calls. Train attendants on special console functions such as alarm monitoring, checking all-trunk busy conditions, releasing trunks, etc.

Training Complete

Module 11: Cutover System

Transfer existing trunks to the new system In accordance with the cutover procedures, any existing trunks are transferred from the old system to the new, keeping service alive to critical stations as necessary. Make call-through tests of all trunks and test each incoming trunk to the console.

Cutover 800/888 numbers The IXC transfers 800/888 trunks to new trunk groups. Where 800/888 service is being moved from an existing to a new trunk group, the IXC transfers the service to the new trunk group. The PBX vendor calls each 800/888 number to ensure that it routes to the proper location.

Transfer stations to the new system

Technicians remove crossconnects from old system ports, and transfer them to the new system. The transfer operation and placing and testing of stations are coordinated so that dial tone is provided ahead of station installers.

Cutover Complete

Module 12: Accept System

Provide user follow-up training

Provide training for users who were absent during the pre-cutover training, and for those to failed to understand clearly the first time.

Provide roving assistance

A group of support people, fully trained on system operation, should be provided by the vendor for the first day to offer assistance to people who have questions about using their telephone systems. These assistants should be provided an easy means, such as a special hat or shirt, for users to identify them.

Work deferred changes

If work was deferred because of the freeze date, work the necessary changes in the PBX.

Inspect PBX installation quality

Inspect all aspects of the installation for acceptable quality. See Appendix A for a quality inspection checklist.

Reprint telephone directory

When extension numbers are being changed in conjunction with a PBX change, the internal directory is revised. For companies using a personal computer console, the database must be prepared or revised.

Check security

Perform all manufacturer's recommended tests on the PBX to ensure that the system has been made as hacker-proof as possible. Maintain a record of all security measures taken. See Appendix B for a checklist of security measures.

Provide initial system database record.

Print a copy of the system database. Create a backup copy of the database on disk or tape and move to off-site storage.

Check system configuration

Download system configuration information from the PBX. Check against the system order for compliance. File information with the PBX.

Inventory equipment

Physically verify that all hardware and major software components that were listed in the equipment order are installed or stored as spares. In lieu of physical inventory, telephone sets can be verified against the master station list.

Provide as-built documents

The vendor provides all system documentation that the RFP and response calls for. At a minimum, as-built documents include an equipment room drawing, system bayface drawings showing what cards are assigned to what slots, a system configuration printout, and a record of trunk and station assignments.

Provide manufacturer's documentation

Verify that all documents that the manufacture furnishes with the system are provided and filed. These should include manuals for system general description, installation, maintenance, and system administration.

Perform initial traffic measurements

As soon as feasible following cutover, traffic data should be collected for at least one week. Verify that trunk quantities are appropriate to provide the desired grade of service. Adjust trunk quantities as required.

Review trouble records

The record of trouble for the first 30 days should be reviewed to verify that the system is meeting performance requirements. If requirements are not met, the acceptance period should be extended until the system is clearly performing to the customer's expectations.

Check carriers' bills

Review carriers bills to detect possible malfunctions of the ARS. For example, long distance showing on the LEC's bill indicates that either the ARS is not functioning correctly or that the IXC has failed to identify all trunks.

Accept system

When all requirements have been met, the customer signs the official acceptance documents, which indicates that all money due is payable and that the warranty period can begin unless warranty was to start on cutover.

Project Complete

Chapter 3

ACD Installation and Upgrade

"Young men are fitter to invent than to judge, fitter for execution than for counsel, and fitter for new projects than for settled business."

Francis Bacon *Of Youth and Age*

Automatic call distribution is supplied with the vast majority of large PBXs, and many smaller ones sold today. ACD products are also available as standalone systems from several manufacturers. Most of the elements of a PBX project also apply to ACD, but if the ACD is more than an incidental application on a PBX, appreciably more planning is required. Since the architecture of standalone ACDs and PBXs are similar, much of the same planning process works for both. This chapter is intended for use with Chapter 2. Some of the tasks relating to PBX installation that also apply to ACD have not been included in this chapter.

Most ACD projects will also need the voice mail and IVR applications presented in Chapters 4 and 5. Many will also need to add the CTI planning process outlined in Chapter 6.

In addition to the project modules of the PBX chapter, other modules are added when you plan an ACD. Call flow is of the utmost importance in designing a call center. A poorly designed call flow confuses callers and wastes money on excessive 800/888 costs. The call flow module discusses the major steps. Chart the steps carefully, and consider the content of announcements because these have a major impact on customer satisfaction.

Another important element of ACD design is providing information to managers and agents about service results and the status of queues. Flashing lights, chimes, reader boards, displays on agents sets, and supervisors' terminals all provide service and workload information. Printed reports are an important source of information for evaluating service results, scheduling staff, and forecasting the workload. Since at least half of the value of an ACD comes from the management information it provides, this is an important part of the planning process.

Trunking also requires close attention. Special call routing services are available from the major IXCs, and should be considered in any call center. Special routing and tie trunks are important for call centers that span multiple time zones. Customer service can often be improved by allocating calls directly from the network.

Finally, an ACD, far more than a PBX, is centered around people. Training requirements are more stringent than in a PBX, and involve instruction not only on telecommunications systems, but also on the supporting data systems. Administrative training is likewise important to enable managers to understand reports and prepare administrative people to make the periodic adjustments that help ensure customer service. Training may also be required on IVR, voice mail, and CTI.

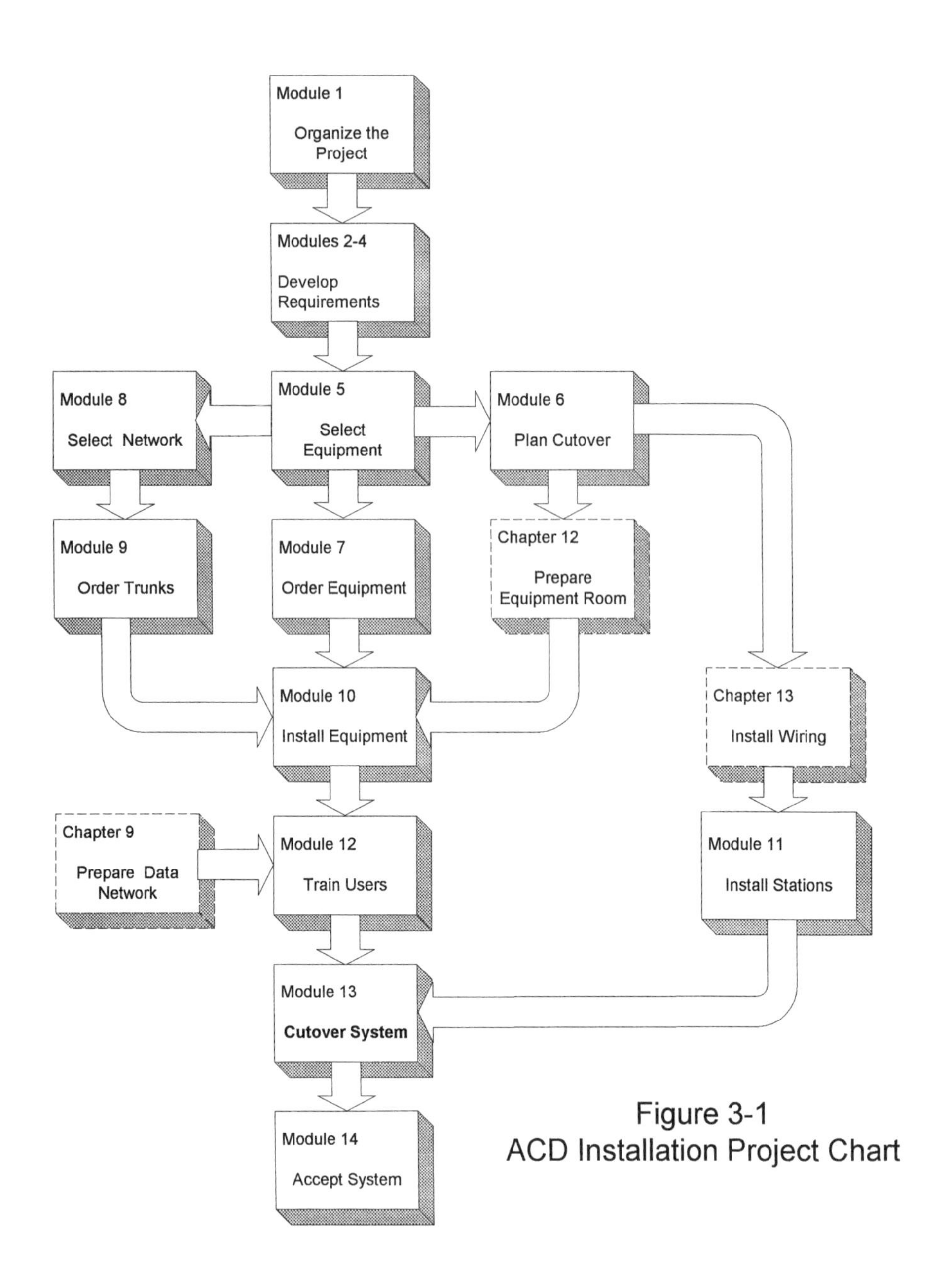

Figure 3-1
ACD Installation Project Chart

Module 1: Organize the Project

Set project objectives

Set objectives for project completion date, budget, impact on customers, and other elements that the project is expected to achieve.

Organize project team

Identify all who are needed in the project. Representatives should be identified from the ACD vendor, which should provide a project manager who can commit the company's resources. The LEC and the IXC should have representatives, although they may not be required at all meetings. If rewiring is involved, the wiring contractor should provide a representative. Identify internal departments that must be represented. Typically, these include the telecommunications, information systems, facilities, human resources and public relations departments. Be certain that roles and responsibilities of all team members are clearly understood and accepted.

Hold kickoff meeting

In the initial project meeting all team members should understand their own and others' responsibilities. The kickoff meeting has the following objectives: assign and accept key responsibilities, communicate objectives and constraints of the project, establish schedules and content of reports and project meetings. Someone should be assigned responsibility for preparing minutes and distributing them within a day or two of the meeting.

Set cutover dates

Determine key dates in the project, including the final cutover dates as well as any interim dates that must be met.

Schedule project team meetings

A complete call center redesign is a lengthy project that requires representatives from several parts of the organization. Project team meetings should be held frequently enough to keep everyone informed of progress, and to resolve issues that arise.

Project Team Organization Complete

Identify project constraints

Identify any constraints that affect completion of the project. These could be such things as force availability, avoiding peak workload periods, space reliability, special promotions, move-in phases, and other such matters that must be taken into account.

Identify call center service objectives

Call center objectives are at the heart of the system design. These objectives relate to such matters as how callers are handled inside the center, speed of answer, number and type of queue announcements, percentage of abandoned calls, etc. Availability of self-service alternatives such as FOD and diagnostics while in queue should be considered. Note the distinction between project objectives, which are handled by the technical staff, and call center objectives, which are controlled by call center management.

Determine sizing variables

Determine the size and type of ACD. Determine such variables as the volume of calls received during peak hours, average talk and wrap-up time, source of calls (800, DID, etc.), number of agent positions, number of agent groups or queues, etc. Project growth through the life of the equipment.

Develop task list

Separate the project into its major components such as facilities, telephone system, data networks, voice network, call center design, etc. Develop a list of tasks for each section. Sequence the tasks and estimate the time each requires. Coordinate tasks between project elements. Identify an individual responsible for each task.

Develop detailed schedule

Integrate tasks from all portions of the project, sequence and link them to produce an overall project schedule. Compress the schedule as necessary to meet the time available. Identify project milestones.

Project Schedule Complete

Module 2: Determine Technical Requirement

Determine agent position requirements

Determine how many agent positions will be required, and how many groups or splits they will be divided into. Determine what types of ports agents will need. Consider the need for analog ports for modems and BRI ports in addition to standard digital ports.

Determine non-agent port requirements

Determine the number of non-agent station ports of different types (analog, digital, BRI) that are required for supervisors, support personnel, etc.

Determine equipment room requirements

Determine the amount of space required for the call center equipment. Include space for switching systems, backup power, multiplexing equipment, server, voice processing equipment, and administrative work space.

Develop agent identification plan

Call centers can be set up with agents identified by position or by ID code. If ID codes are used, agents can occupy any position, and their personal extension number and voice mail follows them. Note that log-in codes may be required for both the ACD and the data network. If possible, the same log-in should be used for both. CTI functions can eliminate the need to log into two separate systems.

Determine ACD networking requirements

Companies with call centers in multiple locations may need to network them so calls from one center can overflow to another. Also, calls can be transferred from one center to another to enable functional specialization. If networking is required, determine the volume and duration of calls. Determine whether systems will be networked with dedicated circuits or if they will use the PSTN.

Determine IXC network routing requirements

Major IXCs offer network routing with their 800/888 services. Calls can be routed on such variables as originating area code, time of day, and day of week. IXCs can also transfer calls and offer IVR and automated attendant functions.

Develop synchronization plan

All devices on a digital network must be synchronized. Devices such as PBXs and routers take their timing from the source closest to the national reference frequency. Other devices on the network slave from the master. Be careful to avoid synchronization loops. Prepare a synchronization plan. Consult an expert if in doubt about plan validity.

Determine earthquake bracing requirements

Depending on the seismic zone in which the system is being installed, determine whether earthquake bracing is required for the switching system.

Determine database requirements

Most call centers require a database in support of agents delivering information over the telephone to customers. If a database does not exist, it will be necessary to determine such issues as what information it must contain, the type and operating system of the hardware, the method of distributing the information to agent positions, and the means by which information will be displayed.

Determine call center data network requirements

New call centers will likely require a data network to distribute information to the agents. Determine the type of network (dumb terminal, LAN, etc.) physical wiring medium, work station requirements. See Chapter 9 on LAN planning and Chapter 13 on wiring plans.

Determine caller identification requirements

Callers can be identified by one or a combination of methods. Methods include ANI, customer-dialed ID codes, DID or 800/888 number, or operator identification. Determine at which stage in the process the caller will be identified and what type of equipment is necessary to do the identification.

Determine agent voice set requirements

Determine what type of telephone set is required by agents. Display is virtually mandatory. The number of feature buttons is determined by the number of functions the agent must perform and by such factors as the number of queues the agent is simultaneously logged onto. Determine requirements for such features as lights or displays showing service information such as calls in queue and length of oldest waiting call. Determine headset requirements including, if necessary, provisions for supervisors to plug in at the position to monitor or talk to a caller. Determine requirements for a serial port connection for CTI applications.

Determine agent data terminal requirements

Determine the type of data devices the agents will use. Determine such issues as amount of memory, monitor type and size, hard disk size, etc.

Determine CTI requirements

Integration between the computer and telephone system is becoming increasingly important in call centers. Determine the need for such functions as screen pop, call data capture, dialing from the database, etc. Determine the objectives of the CTI functions and interface requirements such as the type of database and computer system that must be supported. See Chapter 6 for additional details.

Determine requirements for music or announcement on hold

Determine what will be played during callers' waiting interval. Choices are music on hold, announcements on hold, or a combination of sources.

Determine digital announcement requirements

Determine the number of channels and the amount of recording time required for queue announcements.

Determine IVR requirements

Many call centers required interactive voice response to collect information from callers and to enable callers to retrieve information from the database. Determine how calls will be routed to the IVR. Determine if and how customers will be allowed to transfer from the IVR to a live agent. Determine the number of ports required and the amount of disk storage space needed. Identify programming requirements. See Chapter 5 for additional information on IVR implementation.

Determine voice mail requirements

Determine the degree to which voice mail will be used in the call center. Determine whether voice mail will be offered as a routing step to calls in queue. Determine whether special features such as capture of calling number in voice mail for automatic call back are required. Determine the number of ports and amount of storage required. See Chapter 4 for additional information on voice mail projects.

Determine attendant console requirements Many call centers have no attendant console, while others require all calls to first be answered by an attendant before being routed to a service queue. If consoles are required, determine the quantity and the functions needed.

Determine outbound calling requirements Many call centers combine inbound and outbound functions, while other centers such as telemarketing centers are confined to outbound calling. Determine the type of outbound calling required. Power dialing enables agents to set up a call by dialing from a database. Predictive dialing analyzes the probability of an agent's becoming available and attempts to deliver connected calls with a minimum of wait time for either the agent or the called party. Determine whether functions such as answering machine and voice mail detection are required. For predictive dialers, determine the amount of control managers need over agent wait time. (The more wait time permitted, the lower the probability of the system connecting to a call when no agent is available.) Determine requirements for blending calls between inbound and outbound for the same group of agents.

Determine fax requirements Determine requirements for personal fax lines such as those built into modems, and for common area fax lines. Determine if these will be assigned to analog ports. If so, ensure that the fax number is either compatible with the station numbering plan, that it can be translated to a compatible number within the ACD or that the number can be changed.

Determine modem requirements

Identify all modems that will be assigned to ACD ports. Determine if modems must both send and receive calls. If receiving is required, assign a DID number. Check to ensure that modems are suitably isolated from the network to block hackers.

Develop security requirements

Automated attendants and voice mail must be secured against the possibility of toll thieves dialing through to an outside trunk. IVRs must be secured against unauthorized access to account information. See Chapter 5 for additional details.

Equipment Requirements Complete

Module 3: Determine Management Requirement

Determine supervisory terminal requirements

Most call centers require terminals for supervisors to monitor queue and agent status. Determine the number and type of terminals required. Determine which variables must be monitored. Determine the need for color to show exceptions. Determine the type of connection required or desired to the ACD (RS-232, Ethernet, etc.).

Determine forecasting requirements

Determine what method call center supervisors will use to forecast demand. If manual forecasting will be used, determine the source of information. If mechanized support is required, determine the degree of sophistication required and the computer platform the system should run on.

Determine force scheduling requirements Determine how the work force will be scheduled. If mechanized scheduling support is required, determine sizing variables such as the number of shifts required, the number of agents required to be scheduled per shift, and requirements for scheduling relief and lunch periods. Determine the computer platform the scheduling system should run on.

Determine reporting requirements Call center reporting systems fall into two general categories: systems running in the generic program of the ACD, and those that run on outboard computers. The former generally provide a fixed collection of reports while outboard systems are programmable. Determine how reports will be printed and distributed to users. Determine the requirement for special reports such as all trunks busy, event register, and distribution matrix of answered and abandoned time.

Determine service observing requirements Many call centers monitor a sample of incoming calls for evaluating service. Determine whether silent monitoring is acceptable or whether agents will be informed when calls are being observed. Determine the type of service observing station equipment that is needed.

Determine call recording requirements Some organizations such as brokerages require that all calls be recorded for later verification of customer orders. Determine whether such recording is required, and if so how many simultaneous stations must be recorded.

Determine indexing and playback requirements. Determine for how long and by what methods recordings will be archived.

Management Requirements Complete

Module 4: Determine Call Flow

Determine requirements for personalized agent greetings

To avoid agent fatigue, some ACDs permit agents to record greetings in their own voices. When calls arrive at the agent's position, the greeting is played. Determine whether multiple greetings are required and on which variables such as DNIS numbers.

Determine forced announcement requirements

This feature routes calls to an announcement that must be heard before the call routes to an agent. It is frequently used by utilities to inform callers of known outages, and by other centers to inform callers of special promotions or events such as storms that may cause delay in answering calls. If the feature is required, determine the variety and length of messages required.

Determine need for intelligent queue announcement

A growing trend in call centers is to alter the contents of queue announcements based on conditions. For example, callers can be informed of expected wait time or their position in queue. If such announcements are required, determine what information is to be conveyed to callers and what physical medium (e.g. IVR or conditional routing to a digital announcer) will be used to make the announcements.

Determine need for preferred agent routing	This feature, which may require CTI, routes incoming calls to a preferred agent. Determine what variables will be used for routing (for example, customer preference stored in the database, agent who handled last call, etc.). Determine how callers will be identified and how calls will be routed. Determine whether CTI is required or whether an IVR can be used to route calls.
Determine need for conditional routing	Conditional routing permits the call center to use such variables as time of day, day of week, number of calls in queue, length of oldest waiting call, etc. in routing scripts. Determine which routing variables are required.
Determine need for abandoned call capture	ACDs with this feature can capture the number of callers who hang up without being served and can route the call to an agent to return the call when one becomes available. Determine whether calls will be routed to agents automatically, or reported to supervisors to schedule return calls. Determine how caller identity will be established (caller dialed account code, ANI, etc.).
Determine need for skill-based routing	Determine the need for routing calls to agents that have specific capabilities such as specialized training, language skills, etc. Determine how many routing variables are required. Determine how the customer's needs will be identified ((DNIS or DID number, automated attendant, etc.).

Determine need for home-based agents

Telecommuting is a growing trend in call centers. This feature enables agents to log onto an ACD from home. Determine the quantity of home agents required. Determine method of access (analog line, ISDN, etc.). Determine how they will access the database (ISDN, dedicated trunk, dial-up, etc.). Determine type of telephone set required (analog, proprietary, ISDN, etc.).

Determine incoming call flow

Determine how calls will route from incoming trunks to agents. Identify the major steps in call flow including announcements, routing to special devices such as FOD and IVR, conditional routing, etc. Determine requirements for automated attendant routing. Determine requirements for routing in the IXC network. In networked ACDs, determine routing between systems. Diagram routing in a detailed flow chart.

Determine overflow requirements

Determine under what conditions calls will overflow from one queue to another. Determine interflow (overflow between call centers) requirements. Determine requirements for statistical information to follow call flow.

Call Flow Complete

Develop training requirements

Determine what features of the of the ACD training will be required on. Plan for agent training on telephone set and data terminals. Plan for supervisory training on terminals and interpreting and using reports and for forecasting and scheduling systems. Plan for system administrator training on the switching system, ACD software, data system, and the ACD report writer.

Identify effects on promotional material

Unless the project is transparent to callers, changes in promotional material such as advertisements, brochures, stationery, etc. may be required.

Requirements and Specifications Complete

Module 5: Select Equipment

Prepare RFP

Write a request for proposals or quotations listing all requirements and desirable features of the system. Issue the RFP to selected vendors. Set a due date for responses.

Develop evaluation criteria

Determine how the product will be selected. Price is always a factor, but other factors should be considered such as features included, compatibility with existing equipment, product support capabilities, ease of use, ease of management, and references from users with similar configurations. Review expected

productivity gains from the system, bearing in mind that more than three-fourths of the cost of a call center is personnel related.

RFP Responses Received

Evaluate responses

Vendor responses are reviewed. Non-conforming products are eliminated. Rank the remaining proposals and select the top two or three products for further analysis.

Demonstrate products

The vendor demonstrates the product in an environment similar to the one in which the products will be used. Watch demonstrations of major features. Evaluate the products against the evaluation criteria.

Check references

Discuss with at least three customers their experience with the product. Determine whether the vendor has been sufficiently responsive in problem cases. Get their opinion on ease-of-use in day-to-day operation. Ask whether they would buy the same product again.

Select the product

Choose the winning proposal. Obtain management concurrence. Arrange financing, if appropriate. Negotiate a contract with the vendor.

Equipment Selected

Module 6: Plan the Cutover

Determine agent set installation methods

If an existing ACD is being replaced, determine how new telephone sets will be installed. If spare pairs are available, new telephone sets can be replaced at the agent positions in advance of cutover, which may enable familiarity. If the call center operates 24 hours per day, a plan is needed for migrating agents to other positions so their positions can be converted.

Develop change control procedures

Lack of change control is one of the most common reasons for failure of a cutover. The switch database must be programmed in advance of the cutover. If possible, the vendor programs the new switch to the maximum degree with information dumped from the old database. If change control procedures are not exercised, the risk is high that the new ACD will not reflect the station configuration as it existed in the old one. The most effective way to control changes is to establish a freeze date beyond which new orders will not be processed. If changes must be made in the old switch after the freeze date, a careful record must be kept outside the new switch so the same change can be worked twice: once in the old and in the new switch after cutover. Announce the freeze date well in advance so users have a chance to get their changes made before programming of the new ACD begins. Accumulate requests after the freeze date, and work them after the cutover is complete. If trunks are being reused, control of trunk changes is at least, if not more critical than station changes. Trunks can be added or deleted provided the information is processed in the new switch after the cutover. With changes of all types, it is imperative that the new ACD keep pace with any changes in the old one.

Develop cutover methods

Develop a plan for the transition from an existing ACD. Determine how trunks will be moved. Determine how stations will be converted. If the call center operates 24 hours per day, determine how calls will be routed from one system to the other, and how calls in progress will be maintained. Develop plans for moving self-help functions that normally operate continuously such as FOD, BBS, and RAS.

Determine trunk transition methods Determine how trunks will be moved from the old to the new ACD. Consider the method of preventing a disruption in call flow, and of preventing cutoff of calls in progress. This step is not required when all new trunks are being provided.

Develop QA plan Develop a plan for quality assurance. Include such factors as checking that all stations are included, verifying call routing, verifying announcements, verifying station restrictions, verifying ARS accuracy, verifying compliance with manufacturer's specifications, etc.

Develop station wiring plan Determine the quantity and category of wire to each workstation. Consider both voice and data wiring. See Chapter 13 for additional details.

Develop testing plans Determine how system operation will the tested. Develop plans for testing the switching system, agent positions, voice mail, IVR, and self-help functions (FOD, BBS, RAS). If IXC routing is used develop plans for testing call routing through both the call centers and the network. Develop plans for testing networking functions if center is networked with another ACD.

Prepare floor plans Obtain a floor plan showing the location of all agent positions, supervisory positions, display boards, printers, and switching system and data equipment rooms. Show agent ID, extension number, and telecommunications outlet number on the floor plan or on a separate list cross referenced to the floor plan. Show the location of any special apparatus such as BBS computers, RAS, IVR, supervisory terminals, attendant consoles, and FOD.

Floor Plans Complete

Develop station restriction plans

Develop a plan for restricting selected stations from access to long distance, international calling, 809 area code, 900/976, and other codes that are subject to toll fraud or abuse. Develop a plan for restricting access to features that are subject to toll fraud or abuse such as trunk-to-trunk transfer and off-system forwarding.

Develop trouble reporting procedures

For the first few days following cutover, a help desk will be established and employees will be informed on procedures for reporting trouble. Procedures should also include methods for referring troubles to technicians, maintaining a trouble log, and clearing trouble with the users when clearance is reported to the help desk.

Develop customer publicity plans

When major changes are planned such as introduction of self-help features (IVR, FOD, BBS, etc.) changes in service hours, significant changes in call routing, etc. consider

publicizing the event to regular callers through catalogs, mailers, etc.

Assign station numbers

Determine the station numbering plan and assign extension numbers to the stations, including special stations such as FOD, BBS, etc.

Station Planning Complete

Set station change freeze date

To control the project effectively, a date must be set, beyond which no further station changes will be accepted in the existing switch if the new switch is to obtain database information from the existing switch. If the cutover is to a new ACD, the database must be frozen while the new one is created. Publicize the need for the freeze to all station users and inform them of the last date for placing orders.

Set network change freeze date

Develop a date beyond which no further trunk changes will be ordered. The network freeze date applies to local, IXC, and tie trunks, as well as to any voice mail trunking.

Design system software

The vendor's representative develops coding instructions for all variables in the system software for stations, trunks, agent positions, and queues.

Develop contingency plans

Develop plans for coping with unexpected events that disrupt the project schedule. Consider such factors as delay in equipment arrival, delays in trunk installation, major equipment problems, changes in building occupancy schedule, etc. Also consider

whether an opportunity to advance the schedule may arise.

Cutover Planning Complete

Module 7: Order Equipment

Conduct station reviews Review the requirements of each ACD group or split. Determine the need for personal lines, voice mail coverage, buttons on station sets, and data terminals. Develop configurations for each unique group of agent workstations.

Develop final voice platform configuration In preparation for placing the equipment order, develop the final configuration of voice mail, auto attendant, and IVR. Determine port and storage requirements.

Develop final ACD configuration In preparation for placing the equipment order, develop the final configuration of ACD equipment including trunk configuration, station port configuration, station set types and quantities, MIS configuration, CTI requirements, and software applications.

Order ACD equipment The customer orders the equipment from the vendor, who places the order with the manufacturer.

Order data terminal equipment The customer orders data terminal equipment from the vendor, who orders the equipment from the manufacturer.

Equipment Ordered

Module 8: Select the Network

Determine local trunking requirements Determine quantities of analog, digital, and PRI trunks required from the LEC.

Determine IXC trunk requirements Determine the quantity of trunks required from the IXC. Determine whether trunks are one-way, two-way, or PRI.

Determine 800/888 service requirements Determine requirements for 800/888 trunks. Determine the need for special routing features such as time-of-day, area code, etc. Determine need for network-provided automated attendant or IVR.

Determine tie line requirements For networked ACDs, determine the quantity of tie trunks required to other systems.

Determine WAN requirements Determine the need for wide area data services to support the network. Services may include dedicated analog or digital lines, frame relay, etc. See Chapter 10 for additional details.

Determine DID requirements Determine the quantity of DID numbers required for agents' personal lines, and for calls into special systems such as FOD, BBS, RAS, etc. If coordinated dial plan will be used with another system, consider the effect on station numbering.

Network Design Complete

Module 9: Order Trunks

Order local trunks The customer places order with the LEC for trunks and DID numbers.

Order tie lines Customer orders tie trunks from the IXC or LEC as appropriate. Specify signaling requirements.

Order IXC trunks Customer orders trunks, tie lines, and data WAN circuits from the IXC.

Order 800/888 services Order 800/888 services from the IXC. State the quantity of numbers and DNIS requirements. For business line service, indicate the telephone numbers to which they terminate. Order 800/888 numbers transferred from existing to new trunks, specifying to the IXC the time and date on which the transfer must be made.

Trunk Orders Complete

Module 10: Install Equipment

Prepare equipment room Prepare the equipment room in accordance with the equipment room plan. Refer to Chapter 12 for additional details.

Equipment Room Complete

Install wiring If new wiring is required, or if existing wiring must be modified, complete wiring before station set installation begins. See Chapters 13 and 14 for wiring system planning.

Deliver ACD The manufacture ships and the vendor delivers the ACD to the site. The length of time for this item is the manufacturer's interval from receipt of order to delivery.

Equipment on Site

Unpack and The PBX installer unpacks the cabinets, inventories the system to be sure all parts have

position cabinets	arrived, and puts the cabinets in position. If earthquake bracing is required, it is installed and cabinets are bolted to the floor.
Connect power plant	For AC powered systems the ACD is connected to the UPS supply or directly to AC power. For battery operated systems the rectifier is connected to AC power. Batteries are installed and the rectifier is adjusted to bring the batteries up to full charge. For wet cell systems the cell voltage and specific gravity are recorded after the battery supply is fully charged.
Connect system to power	The system is connected to AC or DC power and to ground. Be certain that the ground meets all manufacturer specifications for tightness of connections, wire gauge, and bonding to power.
Install cable supports	Install cable trays or racking to support cable between the cabinets and between the cabinets and the MDF. Cables must be lashed to the cable rack and separated physically according to the manufacturer's specifications.
Power up system	Power is applied to the system and all of the manufacturer's installation tests are run. The installer checks completion of all items on a master check list .
Load system program tape	The generic program is loaded. System initialization tests are performed and recorded on a master check list. All system diagnostics are run and a record of their completion is maintained.

Test battery backup
AC power is disconnected to the battery or UPS supply and left disconnected long enough to verify the capacity of the system. A record is made of the elapsed time from disconnect until the system is incapable of call processing because of low voltage. For DC systems the bus bar voltage at the system failure point is recorded.

Input system databases
Vendor installs trunk and station database. A complete database record is printed and maintained.

Program ARS
The ARS is programmed according to instructions provided in the ARS plan. All programming must be tested to ensure that calls are directed to the right trunk group, that they overflow to secondary groups when all trunks are busy, and that the correct digits are inserted and deleted.

Install and test trunking
Where new trunking is being added, the LEC and IXC install trunks. Analog CO trunks are tested for dial tone and signaling. Install CSUs on T-1 trunks. LECs and IXCs loop back trunks and test for integrity and error-free performance.

Trunks Installed

Connect trunks to system
The ACD installer connects trunks and verifies operation and hunt sequence. Check all analog DID trunks individually with the central office.

Test network features
Test all features that are included in network orders. These include call-through tests of all trunks, test of features such as call transfer,

call-by-call service selection, trunk hunting, forward on busy or no answer, etc.

Install attendant consoles Connect the attendant console to the switch and test operation of all features. In a multi-console environment, test to be sure that calls are distributed between consoles according to the specifications. Verify that the console attendant sees the proper trunk group designations on the alphanumeric readout.

Install voice mail Turn the power on for voice mail and observe that the system boot-up occurs properly.

Install MIS computer Install the computer that contains the management information system software. Load operating system. Load application software.

Computer Installed

Install MIS printer Install the printer(s) that used for management information. Connect printers to the MIS computer.

Test MIS Test the management information system to ensure that it produces the required reports.

Install CTI software If the center is using any CTI applications, install the software on the server, host computer, work stations, and switching system as appropriate.

Test CTI applications Test CTI software to ensure that it functions properly. See Chapter 6 for additional details.

Perform installation Perform all of the initialization tests required in the manufacturer's manual. Prepare a written check list to verify that all tests were performed

tests and provide the list to the customer.

Testing Complete

Program default greetings

The vendor or the customer must record greetings that the voice mail system requires. This may include dial-by-name instructions, and any greetings that the manufacturer permits the customer to change.

Program voice mailboxes

Voice mail boxes are programmed according to instructions. Use caution in programming any voice mail boxes associated with inactive stations because of the risk of capture by hackers.

Program station restrictions

The classes of service are programmed in accordance with the restriction plan. Restrict the use of certain features that are toll fraud prone. Access to hacker-vulnerable country codes, area codes, 800 numbers, and 900/976 numbers should be restricted in accordance with the plan. Restrictions are tested to ensure that programming is correct.

Program administration terminal security

Determine the administration terminal password. The password should be long enough to make it difficult to hack. Security is improved by having one or more special characters in each password. If a remote terminal security device is being installed, program and connect it. The terminal is connected to a business line. DID ports should be avoided so the terminal can be accessed if the switch is down. If a separate voice mail administration port is involved, the same provisions apply except that the voice mail terminal can be assigned to a DID number.

Connect call accounting system

If a call accounting system is used, connect it to the SMDR port and program it. See Chapter 8 for details in setting up a call accounting system.

Perform final system tests

At this point the switch is installed and ready to cut over. Final system inspections should be performed as described in Appendix A. If time permits, the ACD should be allowed to burn in without traffic for at least a week to screen out any defective components. Any final tests recommended by the manufacturer's installation manual should be performed at this point.

Equipment Installation Complete

Module 11: Install Stations

Prepare station installation floor plans

Mark a set of floor plans for the installation crew to use in placing agent sets. The floor plans must show extension numbers of the agent stations plus any additional ports that are provided for faxes, modems, BRI, non-ACD stations, etc. If extension numbers are being changed the floor plan should be marked with both the old and new numbers, or an old-number-new-number list should be provided.

Obtain agent furniture	If new furniture is being acquired for the call center, the vendor delivers and install it.
Provide staging area	In projects that extend over several days, the vendor will require a lockable space for storing telephone sets and some installation materials. Identify space in the building large enough to contain the equipment and materials. Note that equipment room storage is generally undesirable because of the flammable the and dust-generating nature of packing material.
Designate agent sets	When new sets are being placed, boxes are opened and designation strips put on sets. The room or cubicle number is written on the outside of the box. Sets are grouped by area for delivery at cutover time.
Place agent equipment	After cutover to the new system, the new station equipment can be activated. If sets are in place, they can be tested immediately, but if sets are placed during the cutover period, the set placement process must be coordinated with the process of running crossconnects. Leave user manuals at each station unless they are being provided during the training classes.
Call-through test each station	The agent telephone sets are installed and connected to the network. Call-through test each station. Calls are placed to the personal extension number on each station. Vendor tests call distribution to each station.
Test station coverage path	Each station's coverage path is tested by leaving the station off hook while a call is placed and checking to be sure that the call goes to voice mail, the attendant console, or

other designated coverage location.

Install data terminals Deliver data terminals to the agents' work area, power up, and connect to the network.

Load workstation software Load application software on the agent work stations.

Load database in host computer Load the database in the host computer or server. Load customer or information database. Install any security applications required.

Connect terminals to host Connect data workstations to the network and establish connectivity to the host computer or server.

Test data terminals Test all data devices for access to authorized servers and host computers.

Set up help desk A help disk staffed with trained personnel is set up to operate for the first few days after cutover. The purpose of the help desk is to accept trouble reports from users, clear them if possible by providing telephone assistance, and dispatch trouble when it cannot be cleared over the telephone. The help desk keeps logs of all reports received and cleared. When the quantity of reports drops to an acceptable level, the help desk reverts to normal company operations.

Station Installation Complete

Module 12: Train Users

Determine training Determine the location of the training room for agent training. Determine how the room will be

room location	equipped. Consider training for both agent sets and data devices.
Develop training schedule	Training classes are scheduled and participants notified. Determine whether training will be a single class or a series of classes on different elements of call center operation.
Set up training room	The ACD vendor wires temporary station cables into the training room(s), and connects agent sets. An appropriate number of sets of each type that will be used are installed so users can be trained on the type they have on their desks. Set up the attendant console in a convenient location for training the attendants. If on-site administrative training is being provided by the vendor, provide space for this.
Train agents	Train agents on all call center functions such as logging in and out, how to use status keys, meaning of status lamps and reader boards, call monitoring, how to obtain supervisory assistance, etc. Train agents on terminal usage. Train agents on work functions in the call center.
Train station users	The vendor's trainer delivers classroom training as agreed to in the training plan. Pass out user manuals if they are being provided for during training as opposed to being left at the station at cutover. Inform people of how to reach the help desk.
Train console attendants	Train console attendants on how to answer, transfer, and queue calls. Train attendants on special console functions such as alarm monitoring, checking all-trunk busy conditions, releasing trunks, etc.

Train supervisors Train supervisors on all functions on which agents are trained. Also train on how to use supervisory terminals and reports.

Train system administrators Train system administrators on how to change call flow, announcements, add and remove agents, change report format and schedules, and other such functions provided by the ACD.

Training Complete

Module 13: Cutover the System

Transfer existing trunks to the new system In accordance with the cutover procedures, any existing trunks are transferred from the old system to the new, keeping service alive to critical stations as necessary. Make call-through tests of all trunks and test each incoming trunk to the console.

Cutover 800/888 numbers The IXC transfers 800/888 trunks to new trunk groups. Where 800/888 service is being moved from an existing to a new trunk group, the IXC transfers the service to the new trunk group.

Test 800/888 numbers The ACD vendor calls each 800/888 number to ensure that it routes to the proper location.

Transfer stations to the new system Technicians remove crossconnects from old system ports, and transfer them to the new system.

Cutover Complete

Module 14: Accept System

Provide agent follow-up training	Provide training for agents who were absent during the pre-cutover training, and for those to failed to understand clearly the first time.
Provide roving assistance	A group of support people, fully trained on system operation, should be provided by the vendor for the first day to offer assistance to people who have questions about using their telephone systems. These assistants should be provided an easy means, such as a special hat or shirt, for users to identify them.
Work deferred changes	If work was deferred because of the freeze date, work the necessary changes in the ACD.
Perform quality inspections on ACD	Inspect all aspects of the installation for acceptable quality. See Appendix A for a quality inspection checklist.
Check security	Perform all manufacturer's recommended tests on the ACD to ensure that the system has been made as hacker-proof as possible. Maintain a record of all security measures taken. See Appendix B for a checklist of security measures.
Provide initial system database record.	Print a copy of the system database. Create a backup copy of the database on disk or tape and move to off-site storage.
Check system configuration	Download system configuration information from the ACD. Check against the system order

for compliance. File information with the switching system.

Inventory equipment Physically verify that all hardware and major software components that were listed in the equipment order are installed or stored as spares. In lieu of physical inventory, telephone sets can be verified against the master station list.

Provide as-built documents The vendor provides all system documentation that the RFP and response calls for. At a minimum, as-built documents include an equipment room drawing, system bayface drawings showing what cards are assigned to what slots, a system configuration printout, and a record of trunk and station assignments.

Provide manufacturer's documentation Verify that all documents that the manufacturer furnishes with the system are provided and filed. These should include manuals for MIS system administration, system general description, installation, maintenance, and system administration.

Perform initial traffic measurements As soon as feasible following cutover, traffic data should be collected for at least one week. Verify that trunk quantities are appropriate to provide the desired grade of service. Adjust trunk quantities as required.

Review trouble records The record of trouble for the first 30 days should be reviewed to verify that the system is meeting performance requirements. If requirements are not met, the acceptance period should be extended until the system is clearly performing to the customer's expectations.

Check carriers' bills Review carriers bills to detect possible malfunctions of the ARS. For example, long distance showing on the LEC's bill indicates that either the ARS is not functioning correctly or that the IXC has failed to identify all trunks.

Accept system When all requirements have been met, the customer signs the official acceptance documents, which indicates that all money due is payable and that the warranty period can begin unless warranty was to start on cutover.

Project Complete

Chapter 4
Voice Mail Systems

"Any project not worth doing at all is not worth doing well"

Anonymous

Voice mail is usually installed as part of a PBX project, and is mentioned in Chapter 2 as part of a PBX installation. Voice mail is often added to an existing PBX or installed stand alone. Many of the tasks are the same as those in Chapter 2, but there are enough differences that a separate project plan is appropriate. This section covers voice mail as both a stand alone and as an add-on system, and covers detail that is not included in Chapter 2.

Determine the purpose for which voice mail is being installed. Conventional purposes such as covering the phone, conveying information without simultaneous availability, and providing auto attendant services are common. Many voice mail systems have objectives, however, that affect the size and configuration. Audiotex systems deliver verbal information over the telephone, and may require additional resources. Close evaluation may reveal that an IVR is a better vehicle, particularly if the information exists in a computer database. If the voice mail interfaces an ACD, special functions such as providing personal greetings for agents and capturing calling number for automatic callback may be important. Where multiple locations are involved, voice mail systems often must be networked together using manufacturer's proprietary protocols or a standard such as AMIS.

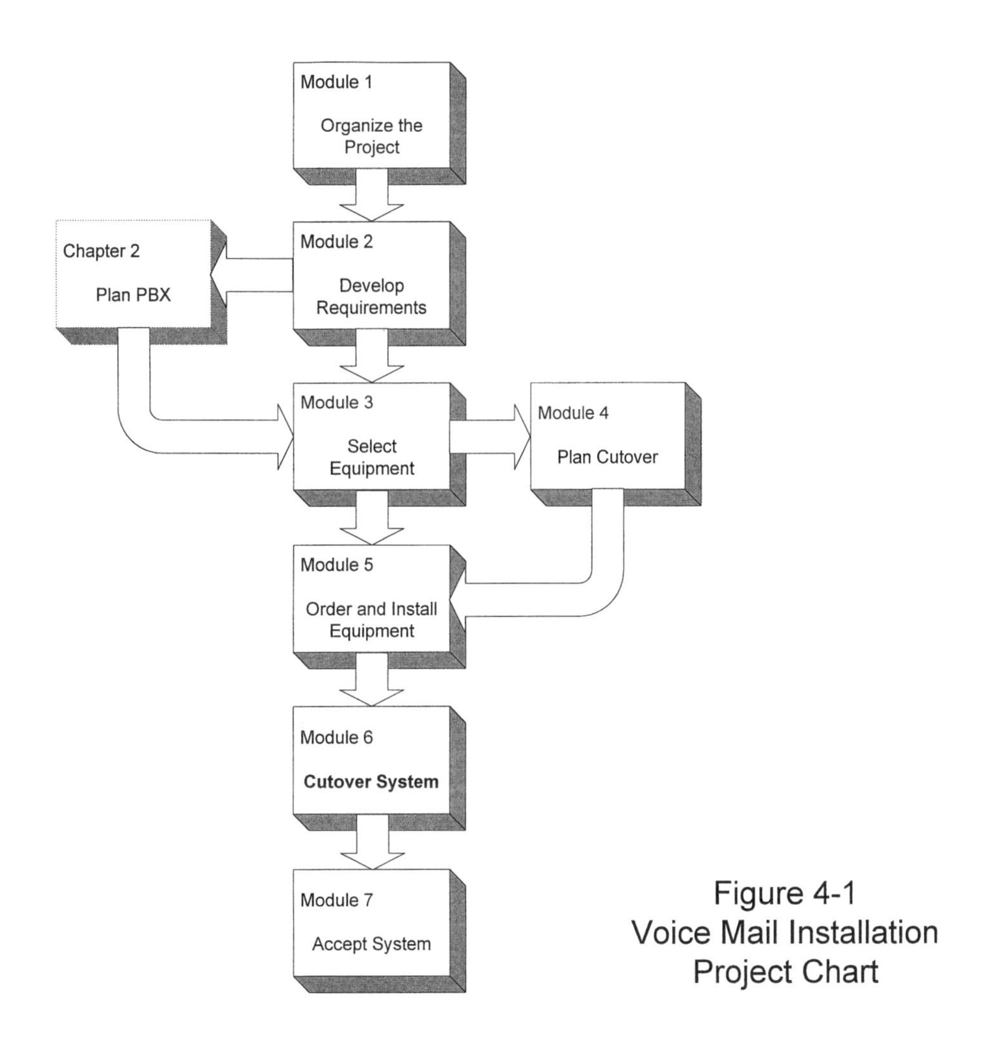

Figure 4-1
Voice Mail Installation
Project Chart

Module 1: Organize the Project

Set project objectives

Determine key objectives for the project. These must include completion date plus dates for any intermediate tasks that may affect target completion dates. Determine budgets for the project.

Organize project team

Identify all who are needed in the project. A representative from the PBX vendor may be required. Identify internal departments that must be represented. Typically, these include the telecommunications, information systems, facilities, and in some projects, human resources and public relations departments. Be certain that roles and responsibilities of all team members are clearly understood and accepted.

Hold kickoff meeting

In the initial project meeting all team members should understand their own and others' responsibilities. The kickoff meeting has the following objectives: assign and accept key responsibilities, communicate objectives and constraints of the project, establish schedules and content of reports and project meetings. Someone should be assigned responsibility for preparing minutes and distributing them within a day or two of the meeting.

Set cutover dates

Determine key dates in the project, including the objective cutover dates as well as any interim dates that must be met.

Schedule project team meetings

Develop a schedule and place for future team meetings. The frequency of meetings depends on the competence of the project team, the complexity of the project, the penalties for missing due dates, and the numbers of people who must be kept informed.

Project Team Organization Complete

Identify constraints

Determine from all project team members any constraints they have with respect to force availability, inability to work certain dates, availability of key personnel, or other factors that may affect project completion. Document all such constraints in the minutes

Develop task list

Develop a detailed list of tasks that must be performed. Assign each task to a responsible individual. Sequence tasks in the order in which they must be performed. Obtain estimates of the amount of time required for each task.

Develop detailed cutover schedule

Identify milestones in the schedule. Set dates for completion of each milestone. Determine the critical path and determine whether the schedule fits within the objective interval. If it does not, determine how to compress the schedule until it fits. Create Gantt and/or PERT charts showing task sequence and schedules, and distribute to all team members.

Project Schedule Complete

Module 2: Develop Requirements

Determine technical

Is a proprietary interface to a particular brand of switching system essential? Ask enough questions to be sure you thoroughly understand

requirements the application. Voice mail systems are sized by the number of ports and hours of storage. As a rule of thumb, average usage is about three minutes per person per day; heavy usage is five minutes. To calculate the number of ports required, assume that 17% of the usage occurs during the busy hour, and calculate ports from standard Erlang B tables.

Determine automated attendant requirements

Identify requirements for automated attendant and audiotex or other applications that increase port and storage requirements. Add these requirements to the standard voice mail for calculating port requirements.

Determine integration requirements

Integrated voice mail systems act in concert with the switching system to light message-waiting lights, escape to a personal attendant, and other features that require the voice mail to integrate closely with the switch. The degree and type of integration required has a major effect on the choice of voice mail system.

Determine outdialing requirements

Many organizations require outdialing to pagers or cell phones from voice mail. Determine how many users require the feature and what level of usage may be required. Note that some voice mail systems require dedicated ports for outdialing, which may increase the port requirements.

Determine interface requirements

Voice mail systems have a variety of different interfaces to the switching system. Some standalone systems have nothing more than an analog port interface, and therefore provide limited integration features. Proprietary voice mail systems provided by the switching system manufacturer are tightly integrated with a processor-to-processor link. Other systems use an external interface device that emulates a display telephone to integrate with a variety of switching systems.

Determine voice mail networking requirements

Voice mail systems can be networked together to enable users to forward messages and send messages to distribution groups. Voice mail systems of the same manufacture can often be networked using proprietary protocols. Systems of different manufacture can be networked using analog or digital AMIS. Determine which type of networking is the most effective for the application. Determine whether networking will be over separate circuits or whether connections will be dialed-up. Estimate the traffic volume.

Determine routing methods

Automated attendant applications or any voice mail functions that are accessed from DID or 800/888 numbers require routing changes in the switching system. For each path into the voice mail determine call flow. Determine how each automated attendant menu is accessed.

Requirements and Specifications Complete

Module 3: Select Equipment

Develop a request for proposals

Develop a document stating requirements and terms of purchase of the voice mail system. Issue the RFP to selected qualified vendors. Set a due date for responses.

Develop evaluation criteria

Determine how the product will be selected. Price is always a factor, but other factors should be considered such as ease of use, effectiveness of integration with the switching system, speed of operations, and references from users with similar configurations.

RFP Responses Received

Review proposals

Review the proposals submitted. Eliminate those that fail to meet requirements. Select the top three or four finalists. Develop a list of questions for proposals that will be considered further.

Demonstrate voice mail systems

Request all finalists to demonstrate their systems. Pay particular attention to the usability of prompts, quality of voice, and operation of integration features. If the system cannot be demonstrated on the same type of PBX it will be used on, obtain a list of contacts of people who have similar configurations and discuss their experience.

Select the system to purchase

Choose the winning proposal, and write a contract covering the terms and conditions of the purchase.

Equipment Selected

Module 4: Plan Cutover

Determine back door access method

Every voice mail system needs an extension number for users to check for messages. A DID number may be assigned for checking from off-site without attendant assistance. Determine whether an 800/888 number is required for remote access.

Develop training plan

Determine how users will be trained on the new system. Systems that are being installed in conjunction with new PBXs provide training as part of the PBX station training, but all add-on and standalone systems require a special training strategy. Determine whether it will be necessary to set up a training room and special training telephones. Determine how people will be trained on security and operational matters such as forced password change.

Develop training materials

Determine what type of printed information such as user manuals will be provided to users. Compile these if not provided by the vendor.

Develop usage policies

To get the most from voice mail, organizations should have policies relating to its use. For example, to prevent phone tag, policies should be set on when it is acceptable to use voice mail to screen calls. Set policies on types of calls that may and may not cover to voice mail. For example, many companies require outside calls to selected groups of people to cover to live attendants, while internal calls may cover to voice mail.

Develop testing plan

Develop a plan for testing all voice mail functions for proper operation. Include a testing plan for trunks or ports, and for routing software.

Develop security plan

Determine how the system will be secured against voice mail hackers capturing mailboxes, and toll thieves using the system to transfer to an outgoing trunk. Develop plans for minimum password length and forced password change if permitted by the system.

Cutover Planning Complete

Module 5: Order and Install Equipment

Order the

Place a purchase order with the selected vendor.

equipment Vendor orders the system from the manufacturer.

Equipment Ordered

Deliver equipment The vendor receives the equipment from the factory and delivers it to the site. Equipment is unboxed and set in the space provided.

Equipment On Site

Power-up equipment The equipment is connected to AC power or UPS, and run through its initialization routines. The vendor records all initialization information.

Configure system Set the system defaults for such variables as password length, number of ineffective attempts permitted, password expiration intervals, defaults for message storage time, classes of service, etc.

Configure networking If voice mail networking is used, configure the software according to the manufacturer's recommendations. If separate circuits are used, connect the circuits to the voice mail system.

Set up security Security measures detailed in the security plan are programmed. Passwords are changed from the default. The PBX and/or the voice mail are programmed to prevent hackers from calling through to trunks, tie lines, and trunk access codes. The maintenance port is secured with hardware or software features to prevent unauthorized access. System defaults for minimum password length and frequency of change are set.

Prepare scripts Prepare all scripts, announcements, automated attendant menus, and other information that must be recorded.

Set up back door access Connect the voice mail to a station port to enable users to check messages remotely. Assign a DID number to the port if appropriate. Connect to an 800/888 number if necessary.

Connect voice mail to switching system The voice mail is connected to the PBX or public network. Ports are tested.

Voice Mail Connected

Record announcements and greetings Record automated attendant menus, announcements, and all other variables that the design calls for.

Test system The vendor and customer conduct tests of the system according to the testing plan. All ports are tested for operation. Remote access methods are tested. All menus are checked to be certain routing software takes calls to the correct destination. All recordings are checked for quality and accuracy. Test outdialing features.

Test voice mail networking Test all supported features to verify proper operation of voice mail networking.

Program user database Users' mailboxes are programmed into the system. Establish classes of service for each user. Program individual box variables that are not set by class of service such as amount of message storage space allocated, identity of the personal attendant, message retention period, or other variables required by the system.

Set up training room If a special room will be required for training users, determine requirements and location. Set up the room and prepare training class material.

Module 6: Cutover System

Train users Deliver user training. Hand out instructional material if provided.

Set up help desk A help desk is set up to assist users who are having difficulty with the new system. The help desk stays in operation for a day or two or until users are no longer experiencing significant problems, after which the trouble process reverts to normal operations.

Cutover system The system is placed in service and tested thoroughly for all functions.

Cutover Complete

Module 7: Accept System

Evaluate service Conduct post-cutover service evaluation. Determine whether the system is meeting its objectives. Review statistics from the voice mail to determine whether its usage is meeting expectations. Follow up on users who may be violating company policies such as keeping the telephone continually forwarded to voice mail.

Accept system When the system has met all of its requirements, it is accepted, and final bills are paid.

Project Complete

Chapter 5
Interactive Voice Response

"Could everything be done twice, everything would be done better"

German proverb

Interactive voice response (IVR) systems have a wealth of applications in most companies. IVRs come in a variety of configurations. Some are self-contained, connect directly to the telephone network, and deliver information contained in the IVR. Other systems connect to a host computer, either directly or through a PBX or ACD, and read information from the computer in response to queries over the network. Some units are connected to ports in a PBX or ACD, where the switching system connects the IVR to the network. In this kind of application it may be necessary to allow callers to transfer from the IVR to agent positions connected to the switch or ACD. Many of the tasks listed in this section will not be required in some projects because of the architecture of the IVR.

Because of the complexity of many projects, you may want to use this chapter in conjunction with Chapters 2, 3, or 4, which discuss PBX, ACD, and voice mail projects. Bear in mind that an IVR is often the initial contact customers have with the company. Unless it is intuitive to use, the risk of offending, confusing, or losing customers is high.

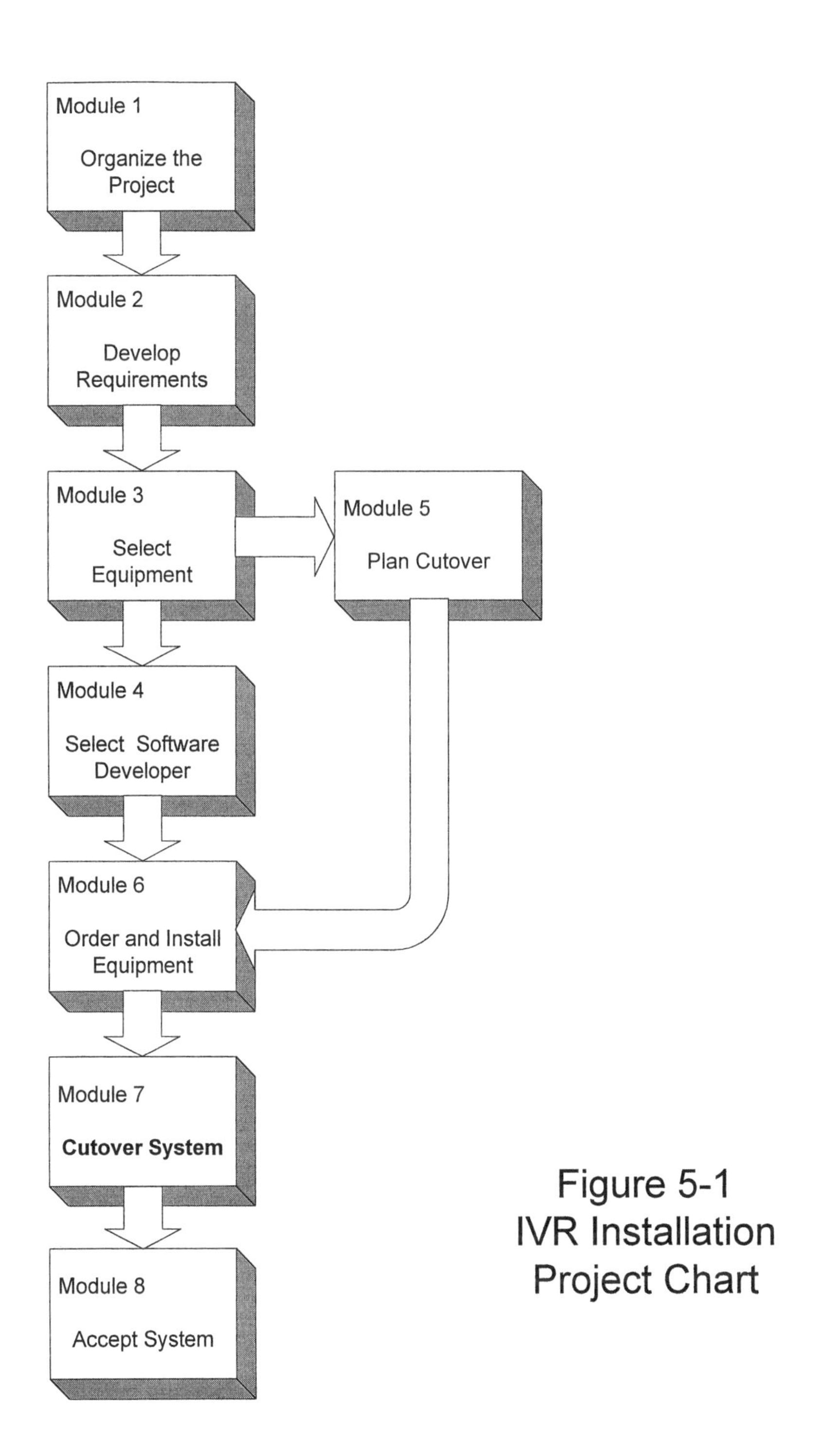

Figure 5-1
IVR Installation
Project Chart

Module 1: Organize the Project

Set project objectives

Determine key objectives for the project. These include completion date plus dates for any intermediate tasks that may affect target completion dates. Determine budgets for the project. Determine customer service objectives in terms of delay time, transfers to and from the IVR, and range of applications that will be installed.

Organize project team

Identify all who are needed in the project. Representatives should be identified from the switching system vendor if the IVR connects through a PBX or standalone ACD. Identify internal departments that must be represented. Typically, these include the customer service, telecommunications, information systems, facilities, and in some projects, human resources and public relations departments. Be certain that roles and responsibilities of all team members are clearly understood and accepted.

Hold kickoff meeting

In the initial project meeting all team members should understand their own and others' responsibilities. The kickoff meeting has the following objectives: assign and accept key responsibilities, and communicate objectives and constraints of the project, establish schedules and content of reports and project meetings. Someone should be assigned responsibility for preparing minutes and distributing them within a day or two of the meeting.

Set cutover dates

Determine key dates in the project, including the final cutover dates as well as any interim dates that must be met.

Schedule project team meetings

Develop a schedule and place for future team meetings. The frequency of meetings depends on the competence of the project team, the complexity of the project, the penalties for missing due dates, and the numbers of people who must be kept informed.

Project Team Organization Complete

Identify constraints

Determine from all project team members any constraints they have with respect to force availability, inability to work certain dates, availability of key personnel, or other factors that may affect project completion. Document all such constraints in the minutes.

Develop task list

Develop a detailed list of tasks that must be performed. Assign each task to a responsible individual. Sequence tasks in the order in which they must be performed. Obtain estimates of the amount of time required for each task.

Develop detailed cutover schedule

Identify milestones in the schedule. Set dates for completion of each milestone. Compute the critical path and determine whether the schedule fits within the objective interval. If it does not, determine how to compress the schedule until it fits. Create Gantt and/or PERT charts showing task sequence and schedules, and distribute to all team members.

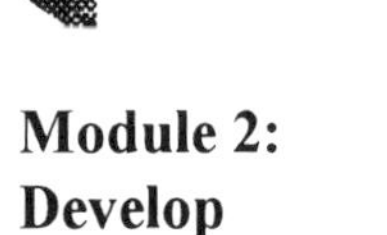

Project Schedule Complete

Module 2: Develop Requirements

Determine requirements

Identify the applications for which the IVR will be used. Determine the interfaces the system must have to the telephone network and to the host computer. If the IVR will contain the information database, determine storage requirements. Determine the number of ports needed. Determine whether ports should be connected through an ACD to enable callers to queue when all ports are busy.

Determine application software requirements

Determine how application software will be obtained. Will you use pre-programmed applications that run on the IVR itself? Does application software run on a host computer? What custom software will be required, and who will write it? If package software is required, identify the packages involved and the purchase procedures.

Develop database requirements

Determine the structure of the database that will be interfaced by the IVR. Determine what functions will be available through voice response. Determine what keystrokes or commands the IVR must send to the database to retrieve each function to which access is permitted.

Develop call flow diagram

Develop a flow chart to show how calls flow into and through the IVR. Show each call source, such as DID, transfer from attendant or service representative, 800/888 number, etc. Determine what the caller dials to reach the source, and over what trunks or ports the call reaches the IVR. Chart the process through the telephone network to the IVR and into the connected computers or self-contained IVR database.

Develop physical network requirements

Determine what physical network will be used to connect the IVR to external devices. On the front end toward the voice network the connection will likely be over cable facilities or T-1. On the back end toward the computer, the connection may be LAN, EIA-232, or proprietary computer interface.

Develop security provisions

Develop the process that will be used to prevent access from the IVR to restricted files. Develop password requirements. Develop a process for enabling callers to establish their initial passwords, and for changing passwords.

Requirements and Specifications Complete

Module 3: Select Equipment

Prepare request for proposals for IVR

Prepare an RFP, bid request, or other procurement document to state requirements for the IVR. Set a due date for responses. State all functions the IVR must perform, and interfaces to network equipment. State capacity in ports and storage space.

Develop evaluation criteria

Determine how the IVR will be selected. Price is one factor, but other factors should be considered such as ease of use, effectiveness of integration with the switching system, speed of operations, and references from users with similar configurations.

Issue RFP

Issue the RFP to potential sources. If custom software is required, determine potential sources, determine whether the IVR vendor will be required to furnish the software or collaborate with a third-party developer.

RFP Responses Received

Review proposals

Vendors' proposals are reviewed, compared to the requirements and specifications and compared to criteria set by the evaluation team. Narrow the proposals down to the top two or three finalists.

Demonstrate proposed IVRs

The finalist vendors demonstrate their products in operations similar to the environment in which the system will work.

Check references

Check references from users who have had experience with the system in a similar application.

Select equipment

The winning proposal is selected and contenders notified. Vendor joins the cutover team and participates in subsequent planning.

Equipment Selected

Module 4: Select Software Developer

Prepare programming specifications

Where custom software is required, develop specifications for programming required to make the system work in your application. Obtain record formats, physical and logical interface formats with both the switching system and the host computer, and a complete description of the expected results. Include due dates and a complete schedule of major events. Detail the results the developer is expected to deliver.

Issue programming RFP

If custom software is required, issue the programming specifications to developers who will be invited to prepare proposals.

Select software developer

Review proposals from developers and select a firm to write the interface between the IVR and the database.

Software Developer Selected

Develop software test procedures

Develop a method of testing all possible functions on the IVR. Test every menu choice, and determine how the system functions with unexpected operations such as the caller hanging up, dialing invalid menu choices, dialing partial digits, etc. Devise tests for all potential attempts to defraud the system. Devise tests for invalid password entry.

Module 5: Plan Cutover

Plan the transition

If the system will replace an existing IVR, develop a step-by-step process for routing calls into the new system and away from the old. Consider how calls-in-progress will be handled, or if calls into the old IVR will be blocked for a time to clear the system before cutover.

Provide space for IVR

Select the IVR location. Consider any lead limitations between the IVR and the switching system and host computer.

Develop IVR scripts Develop the menus that callers will hear when accessing the IVR. Develop the voice scripts that will be recorded.

Develop routing instructions If the IVR is used to route calls within the switching system, determine how calls are to route, and develop instructions for the coder.

Record scripts Select someone to read the various scripts into the IVR. Record the scripts and test them with selected testers to ensure that callers will understand them.

Develop security provisions Be certain that switch ports have a restricted class of service to prevent callers from dialing through the PBX. Develop password structure.

Develop testing procedures Consider all the testing steps that must be taken before system is cutover. Procedures must involve testing of the equipment and software as a total system. Numerous test calls involving all combinations that can be thought of must be made.

Develop IVR monitoring process Develop a process for monitoring how well the IVR achieves its service delivery objectives. Determine how to evaluate port utilization. Determine how to evaluate customer satisfaction with the IVR. Determine how to find out whether menus are working successfully and whether customers are getting their needs satisfied. Evaluate the effects on office productivity in terms of reduced agent workload.

Prepare promotional material Develop material to inform callers of the new IVR. Consider prizes or awards for using the system. For IVRs associated with an ACD, consider the use of forced announcements or automated attendant choices to make callers aware of the IVR. Where callers are accessing secure databases, devise a process for getting them passwords and access instructions well in advance of cutover. Prior to cutover, consider promotional announcements to generate enthusiasm about the system. Publicize its benefits such as 24-hour per day service. Consider setting up a demonstration unit for walk-in customers so service representatives can demonstrate the system.

Cutover Planning Complete

Module 6: Order and Install Equipment

Develop final equipment configuration

The customer and vendor develop the final equipment configuration including port quantities and types, storage, ancillary software, etc.

Order IVR

Customer orders the IVR from the vendor. Vendor orders IVR from the manufacturer.

Equipment Ordered

Deliver IVR

The IVR is shipped from the factory and delivered to the site. Unbox and install in the space provided.

Equipment On Site

Power up the IVR	Connect the IVR to AC power or UPS, turn it on, and run through the boot-up process. Load the operating system software.
Load application software	Load the application software in the IVR and test in accordance with test procedures that were developed.
Order trunks	If the IVR connects directly to the public network, order the necessary trunks. If the IVR connects to a PBX or ACD, determine whether the system has sufficient ports to support the IVR. If not, obtain ports.
Install trunks	Connect the IVR to the network. Test connections to the public network or PBX .
Record scripts	Record all voice scripts and announcements in the IVR.
Mail promotional materials	Mail announcement information promoting use of the IVR to customers and other potential callers.
Test the IVR	Test all functions of the IVR system, including hardware and software together in accordance with the testing plan. Check all security provisions to ensure that they have been properly installed.
Install routing	Change the routing tables in the switching system to route calls through the IVR. Program any automated attendant announcements to route calls into the IVR.

Installation Complete

Module 7: Cutover System

Train users

If attendants or customer service representatives are to transfer callers into the IVR for certain operations, train them on how to recognize the application, what to say to the caller, and how to make the physical connection.

Place the IVR in service

The IVR is turned up for service. Monitor the system closely for the first few days to be sure it is meeting expectations.

Cutover Complete

Module 8: Accept System

Evaluate service

Conduct post-cutover service evaluation. Determine whether the system is meeting its objectives. Review statistics from the IVR to determine whether its usage is meeting expectations. If possible, interview customers to determine their satisfaction with the system.

Accept system

When the system has met all of its requirements, it is accepted, and final bills are paid.

Project Complete

Chapter 6
Computer-Telephony Integration

"For time ye lost may nought recovered be"

Geoffrey Chaucer *The Book of the Duchess* 1369

Computer-telephony integration is in its infancy, and few vendors have extensive experience in implementing it. Unlike PBX and LAN implementation, which vendors have done repeatedly, many vendors will be venturing into one of their first applications of CTI. Moreover, CTI applications are customized to a much greater degree than other telecommunications applications. In many ways, implementing a CTI application is akin to custom software development, in which the coordinated efforts of several diverse groups are required. The plan, therefore, resembles custom software planning more than it does telecommunications systems implementation.

Because of the lack of uniformity in CTI applications, many of the tasks outlined in this chapter are broader in scope than the tasks in other telecommunications projects. The project manager, therefore, has a more detailed and exacting role than in most telecommunications projects. The customer's expectations must be defined in excruciating details because, as with all software projects, the success or lack thereof is determined by the degree of attention to detail.

This chapter is designed be used in conjunction with other projects, which will usually be ACD or PBX, and in some cases local and wide area networks. This chapter assumes that the switching system exists, and that CTI is being added. If a new switching system is required, use this chapter in conjunction with Chapter 2 or 3. The unique modules are divided into two paths: computer hardware and software, and switch hardware and software. Of the two, the computer implementation is by far the most detailed and difficult, and involves the least amount of off-the-shelf software. Nearly all of the software in many CTI projects is custom-developed.

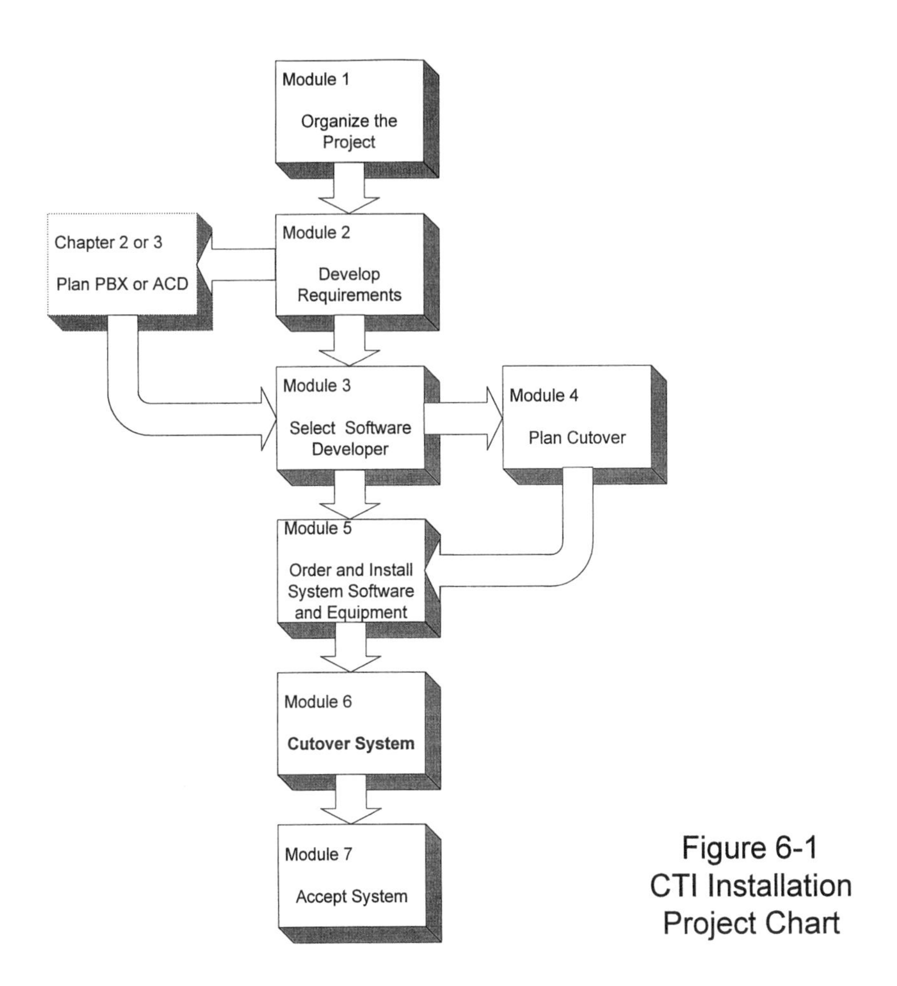

Figure 6-1
CTI Installation Project Chart

Module 1: Organize the Project

Set project objectives

Determine key objectives for the project. These must include completion date plus dates for any intermediate tasks that may affect target completion dates. Determine budgets for the project. Determine exactly what CTI is expected to accomplish, and the company's motivation for installing it.

Organize project team

Identify all who are needed in the project. Representatives should be identified from the switching system vendor. Identify internal departments that must be represented. Typically, these include the customer service, telecommunications, information systems, facilities, and in some projects, human resources and public relations departments. Be certain that roles and responsibilities of all team members are clearly understood and accepted.

Hold kickoff meeting

In the initial project meeting all team members should understand their own and others' responsibilities. The kickoff meeting has the following objectives: develop task lists and sequence tasks, assign and accept key responsibilities, communicate objectives and constraints of the project, establish schedules and content of reports and project meetings. Someone should be assigned responsibility for preparing minutes and distributing them within a day or two of the meeting.

Set cutover dates Determine key dates in the project, including the final cutover dates as well as any interim dates that must be met.

Schedule project team meetings Develop a schedule and place for future team meetings. The frequency of meetings depends on the competence of the project team, the complexity of the project, the penalties for missing due dates, and the numbers of people who must be kept informed.

Project Team Organization Complete

Identify constraints Determine from all project team members any constraints they have with respect to force availability, inability to work certain dates, availability of key personnel, or other factors that may affect project completion. Document all such constraints in the minutes

Develop task list Develop a detailed list of tasks that must be performed. Assign each task to a responsible individual. Sequence tasks in the order in which they must be performed. Obtain estimates of the amount of time required for each task.

Develop detailed cutover schedule Identify milestones in the schedule. Set dates for completion of each milestone. Determine the critical path and determine whether the schedule fits within the objective interval. If it does not, determine how to compress the schedule until it fits. Create Gantt and/or PERT charts showing task sequence and schedules, and distribute to all team members.

Project Schedule Complete

Module 2: Develop Requirements

Define the applications

Determine what CTI applications are required. If appropriate, study the application for cost justification. Principal applications include screen pop, out-dialing from a database, and computer-directed call routing.

Evaluate database requirements

Determine whether the application can be satisfied by the existing customer database and custom programming, or whether packaged applications can be purchased and applied with little or no modification.

Determine switching system requirements

Determine what hardware and software applications are required for the PBX or ACD. Obtain costs and manufacturer's lead-time requirements.

Determine host computer hardware requirements

Determine the hardware and software applications needed (e. g. TSAPI) on the host computer or server. Determine what type of interface card is required. Obtain ordering and cost information from the switch vendor.

Prepare programming specifications

Where custom software is required, develop specifications for programming that is required to make the system work in your application. Obtain record formats, physical and logical interface formats with both the switching system and the host computer, and a complete description of the expected results. Include due dates and a complete schedule of major events. Detail the results the developer is expected to deliver.

Software Requirements and Specifications Complete

Module 3: Select Software Developer

Prepare software RFP

Prepare an RFP, bid request, or other procurement document to state requirements for the software required. Include the requirements and specifications document. Set a due date for responses. State all functions the CTI application must perform, and interfaces to network equipment. Issue the RFP to potential developers.

Develop evaluation criteria

Determine how the system software developer will be selected. Factors to consider include the developer's track record in preparing similar systems, ability to complete the project on time, and references from satisfied customers.

RFP Responses Received

Review proposals

Vendors' proposals are reviewed, compared to the requirements and specifications, and compared to criteria set by the evaluation team. Narrow the proposals down to the top two or three finalists.

Check references Check references from users who have had experience with CTI programmed by the developers in similar applications.

Select software developer Select the software developer based on demonstrated ability to deliver quality software in the required time frame.

Software Developer Selected

Module 4: Plan Cutover

Plan the transition Determine how the transition from the old method to the new will be accomplished. Will the application be provided to all customer service agents simultaneously, or will it be tried on one group? Are routing changes required in the switching system? Are new scripts required?

Prepare software test plan The customer and the software developer create testing methods for evaluating compliance with the specifications. Determine critical test dates and milestones that must be met by a those dates. Determine how test media will be developed. Determine how software quality will be assessed.

Determine hardware requirements Determine what hardware changes must be made to the PBX/ACD and to the host computer system. Determine what types of interface cards must be ordered. Obtain specific lists and price quotations from the vendor.

Determine physical connectivity requirements

Determine specifically how the switching system and host computer will be connected. Determine the type of wire required and how it will be routed. Obtain specific lists and price quotations from the vendor.

Determine host computer software requirements

Determine what software packages or modules such as TSAPI or TAPI must be added to host computers, servers, and/or workstations. Obtain specific lists and price quotations from the vendor.

Determine switching system software requirements

Determine what software modules must be added to the switching system to make it compatible with CTI. Obtain specific lists and price quotations from the vendor.

Cutover Planning Complete

Module 5: Order and Install System Software and Equipment

Order server upgrades

The customer orders the hardware and software needed to make the server or host computer compatible with CTI.

Order switch upgrades

The customer orders the hardware and software to make the switch CTI compatible.

Equipment Ordered

Deliver server upgrades The manufacturer or vendor delivers the hardware and software needed to upgrade the server to CTI compatibility.

Deliver switch upgrades The manufacturer or vendor delivers the switch hardware and software to the customer.

Install switch upgrades The vendor installs switch hardware and software and performs the manufacturers installation tests.

Install server upgrades The vendor installs the hardware and software to make the server or host computer CTI compatible.

Equipment Upgrades Complete

Connect switch and server The switch and server or host computer are physically connected. Server and switch vendors perform manufacture-recommended tests to verify interoperability of the switch and server or host computer.

Load application software Load the application software and test in accordance with test procedures that were developed.

Beta test software The software is loaded on operational equipment and tested for compliance for with the specifications.

Beta Testing Complete

Conduct software acceptance review

Following completion of beta testing, the software developer completes the revisions to the software. Customer reviews software operation and conditionally accepts it pending implementation.

Document operating instructions

The vendor completes user and program documentation and turns it over to the customer.

Load application software on workstations

Application software is loaded on workstations and tested with the server or host computer.

Test CTI applications

All CTI applications are tested in accordance with the testing plan. All functions are tested to ensure that they meet the specifications. Any discrepancies are documented and referred to the developer for clearance.

Modify software as needed

The software developer modifies the software as necessary to fulfill the specifications and to incorporate customer-requested modifications.

Software complete

Module 6: Cutover System

Train users

Train customer service agents or other users on the use of the CTI application. Provide them with user instructions.

Place the CTI application in service

The CTI application is turned up for service. Depending on the transition plan, conduct a field trial with one group or split to determine that all is working properly. If necessary, get the software developer to fine-tune the code. Monitor the system closely for the first few days to be sure it is meeting expectations.

Cutover Complete

Module 7: Accept System

Evaluate service

Conduct post-cutover service evaluation. Determine whether the system is meeting its objectives. Review statistics to determine whether its usage is meeting expectations. Interview users to determine their satisfaction with the system. Check to see if it is meeting its productivity improvement objectives.

Accept system

When all discrepancies have been resolved and the system complies with all requirements, customer accepts it and pays final bills.

Project Complete

Chapter 7
Remote Access Servers

"No one follows the intent of plans as carefully as their authors, and no one resents plans as much as those who inherit them from inferiors, or less experienced personnel."

Robert D. Gilbreath *Winning at Project Management*

Remote access servers are increasingly employed by companies for a variety of purposes. They may link branch offices to headquarters, in which case the RAS automatically dials a connection when it has data to transfer. Servers may enable customers to dial into the company's network to download files or upload orders. Another application is to enable employees who work at home at least part of the time to dial into the company's network to transfer files and access e-mail. It is important that the application be clearly understood at the outset of the project.

Remote access servers bridge the gap between the telephone network and the corporate data network. Therefore, a combination of voice an data networking skills are required to complete an RAS project. This chapter offers suggestions for selecting remote access servers, which come in a variety of configurations that must be matched to the application.

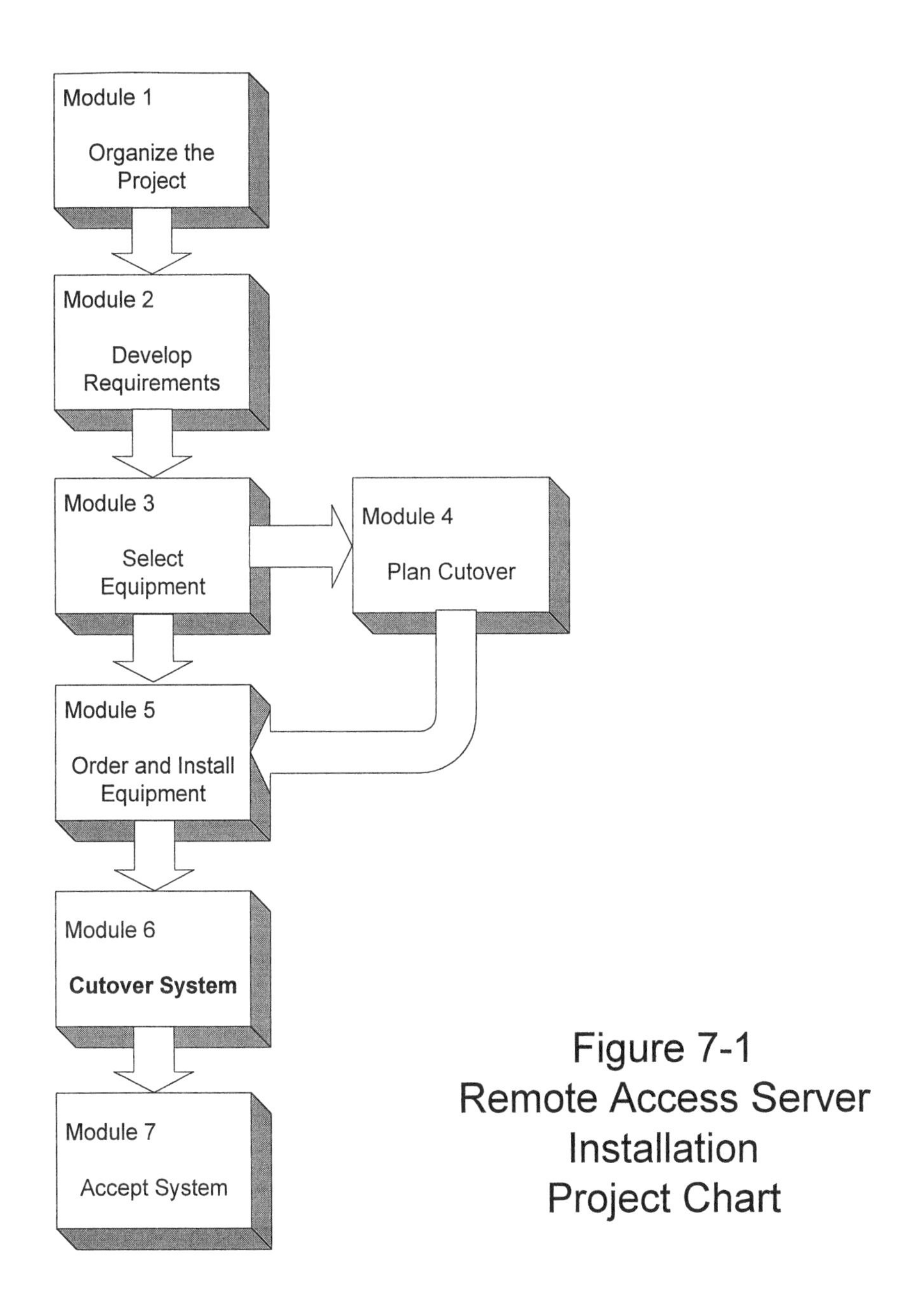

Figure 7-1
Remote Access Server Installation Project Chart

Module 1: Organize the Project

Set project objectives

Determine key objectives for the project. These should include completion date plus dates for any intermediate tasks that may affect target completion dates. Determine budgets for the project. Determine the purpose of the remote access system; for example, to enable telecommuting, link branch offices to headquarters, permit customers to retrieve information from company database, etc. Determine expected growth requirements.

Organize project team

Identify all who are needed in the project, including the vendor providing the server. Identify internal departments that must be represented. Typically, these include the telecommunications, information systems, facilities, and in some projects, human resources and public relations. Be certain that roles and responsibilities of all team members are clearly understood and accepted.

Hold kickoff meeting

In the initial project meeting all team members should understand their own and others' responsibilities. The kickoff meeting has the following objectives: assign and accept key responsibilities, communicate objectives and constraints of the project, establish schedules and content of reports and project meetings. Someone should be assigned responsibility for preparing minutes and distributing them within a day or two of the meeting.

Set cutover dates

Determine key dates in the project, including the objective cutover dates as well as any interim dates that must be met.

Schedule project team meetings

Develop a schedule and place for future team meetings. The frequency of meetings depends on the competence of the project team, the complexity of the project, the penalties for missing due dates, and the numbers of people who must be kept informed.

Project Team Organization Complete

Identify constraints

Determine from all project team members any constraints they have with respect to force availability, inability to work certain dates, availability of key personnel, or other factors that may affect project completion. Document all such constraints in the minutes

Develop task list

Develop a detailed list of tasks that must be performed. Assign each task to a responsible individual. Sequence tasks in the order in which they must be performed. Obtain estimates of the amount of time required for each task.

Develop detailed cutover schedule

Identify milestones in the schedule. Set dates for completion of each milestone. Determine the critical path and determine whether the schedule fits within the objective interval. If it does not, determine how to compress the schedule until it fits. Create Gantt and/or PERT charts showing task sequence and schedules, and distribute to all team members.

Project Schedule Complete

Module 2: Develop Requirements

Determine the purpose of the remote access device

Remote access servers may be used to provide occasional dial-up access to the corporate LAN, to link remote office LANs to the central site, to provide customers with access to a database, or for any number of other purposes. A clear understanding of the purpose must be conveyed to provide the appropriate equipment and configure it properly.

Develop technical requirements

Determine how many dial-up ports are required. Determine whether the dial-up ports will be analog or ISDN. Determine if dial-out modem pool capability is required. Determine the required in-service date.

Determine the platforms system must support

Not all remote access servers are capable of supporting both Macintosh and IBM platforms. Some applications require dumb terminal support in addition to remote LAN connection.

Determine applications that will run remotely

The applications that remote clients use will have a large effect on the choice of remote access system. For example, in a database application, if the database management system runs on the client device the database may have to pass across a dial-up connection to do a file search. Remote access devices used for linking LANs across a dial-up circuit require software that determines when the buffer has reached the point that a connection should be established.

Determine protocol requirements

Determine what protocols (such as TCP/IP, IPX, AppleTalk, MLPPP etc.) must be supported. Determine the protocol and speed of the LAN side of the connection.

Determine mode of remote operation

Remote access servers generally fall into two categories: remote node and remote control. A remote node server requires a proprietary client software application that communicates with the server. After the connection is established the remote node functions like it was a computer attached directly to the LAN. Remote control software enables the remote device to take over the keyboard of a computer connected directly to the LAN.

Determine throughput requirements

Remote access servers are limited by the speed of the dial-up circuit. Determine whether large file transfers or Internet access require ISDN connections or whether the lower speed of analog connections is appropriate. If connections are long distance, consider the additional per-minute cost of switched digital service.

Determine security requirements

By their nature, remote access servers present a security hazard because they permit outsiders to access the company's network from dial-up ports. Determine the need for a firewall to protect the LAN. Determine the need for logs of call activity and length of connections. Determine need for dial-back security. Be certain the system supports an authentication protocol such as PAP or CHAP. Review security provisions in the attached network devices such as servers. Determine how many unsuccessful attempts will be permitted. Determine how hackers will be blocked out and traced.

Determine management requirements

Determine whether the server must support SNMP. Determine requirements for locating and removing defective trunks or modems from service. Determine the need for dial-up access to the server's maintenance port, and for remote reboot.

Determine call monitoring requirements

Determine what variables such as start and stop time of individual calls, login failures, event logging, and user tracking are required.

Requirements and Specifications Complete

Module 3: Select Equipment

Prepare request for proposals

Document a complete list of requirements and specifications, and enter them in a request for proposals or request for quotation.

Issue RFP

Issue the RFP to selected vendors. Set a due date for responses.

RFP Responses Received

Review proposals

Vendor proposals are reviewed. Nonconforming proposals are eliminated. Remaining proposals are ranked and finalists are selected.

Demonstrate products Finalists demonstrate their products in an environment approximating the proposed working environment.

Select product Choose the winning proposal from among those submitted. Negotiate a purchase agreement with the vendor.

Equipment Selected

Module 4: Plan Cutover

Develop training plan Develop a plan for training remote users. If specific applications are being introduced with the RAS, determine how users will be trained on the application as well as the access method. Determine how software will be distributed to users and how they will be trained on its use.

Determine addressing requirements When users dial in for connection to a network that uses IP addressing, some method must be provided for temporarily assigning an IP address for the session. DHCP is the usual method. Determine how dynamic addresses will be assigned, and ensure that the necessary block of addresses is available.

Determine equipment room requirements Determine the space required for the equipment. Choose a location and provide the necessary relay rack or shelf space.

Determine type of network connections Remote access servers can be connected to the public network via analog, BRI, or PRI connections, or to analog or BRI ports on a PBX or hybrid. Determine whether

800/888 access is required. Determine the number of ports and/or PSTN circuits required.

Determine wiring requirements

Determine the type of wire needed for both the LAN and the dial-up port side of the system. Arrange to place additional wiring if needed.

Determine application software requirements

Determine whether additional application software such as client packages are required. If so, determine what packages are needed and in what quantity.

Cutover Planning Complete

Module 5: Order and Install Equipment

Place equipment orders

The customer orders the equipment from the vendor, who in turn orders it from the manufacturer.

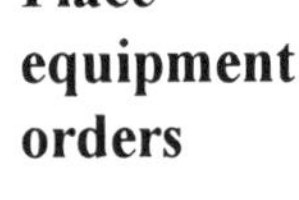

Equipment Ordered

Order application software

Place orders for any application software required.

Order necessary dial ports

For services connected to the public network, order the necessary trunks. For service connected to a PBX, ensure that sufficient analog or BRI ports are available. If new 800/888 numbers are required, order them from the carrier. If the server will add a significant load to DID trunks, check that sufficient capacity is available.

Deliver equipment

The vendor delivers the equipment to the site, unboxes it, and sets it up in the designated location.

Equipment On Site

Power-up equipment

The equipment is connected to AC power or UPS, and run through its initialization routines. The vendor records all initialization information.

Configure system

Set the system defaults for such variables as password length, number of ineffective attempts permitted, password expiration intervals, and other operational and security matters specific to the equipment.

Install software on server

Install the operating system software in the RAS. Assign IP addresses if necessary.

Connect server to switching system

The remote access server is connected to the PBX or PSTN. Ports are tested.

Connect server to LAN

The RAS is connected to the local area network. LAN setup is checked to ensure that appropriate rights and permissions are in place.

Remote Access Server Connected

Test server

Perform all manufacturer's recommended tests. Test security to determine whether sensitive data has been sufficiently isolated from remote access.

Program user database Users' accounts are programmed into the system. Passwords are installed. If dial-back security is used, valid telephone number assignments are entered in the system.

Set up training room If a special room will be required for training users, determine requirements and location. Set up the room and prepare training class material.

Set up remote software The software to enable remote computers to access the server is set up.

Installation Complete

Module 6: Cutover System

Train users Deliver user training. Hand out instructional material if provided. Hand out client software and instruct users how to install on their systems and use it.

Cutover system The server is placed in service and tested thoroughly for all functions.

Set up help desk For the first few weeks after cutover a help desk should be available to assist users who are having problems with remote access.

Cutover Complete

Module 7: Accept System

Evaluate service — Conduct post-cutover service evaluation. Determine whether the system is meeting its objectives.

Accept system — When the system has met all of its requirements, it is accepted, and final bills are paid.

Project Complete

Chapter 8
Call Accounting Systems

"One look is worth a hundred reports"

General George S. Patton

Many organizations that need a call accounting system purchase it in conjunction with a PBX, and simply accept whatever product the PBX vendor offers. This approach may be a mistake. The market offers dozens of call accounting systems, and they vary widely in ease of use, and perhaps more important, in the quality of support provided. We assume in this project that the first step will be to select the product to purchase, after which installation begins.

Call accounting systems range from small, PC-based systems that are used for distributing costs among departments and preparing detailed reports by user, to high-end systems that include telemangement features. These features, which may include trouble reporting, work orders, inventory, and wire management, are available as add-ons to some products. This chapter includes elements of a telemanagement conversion in addition to call accounting. Those elements that will not be part of your project can be eliminated.

Installing call accounting systems is more complex than loading most PC programs. An accurate database is needed because any call records that flow from the PBX's SMDR port must find a matching record in the call accounting system; otherwise they will be lumped into a collection of unidentified calls. In multi-PBX environment, there are buffers to install, polling PCs to set up, and myriad details to coordinate.

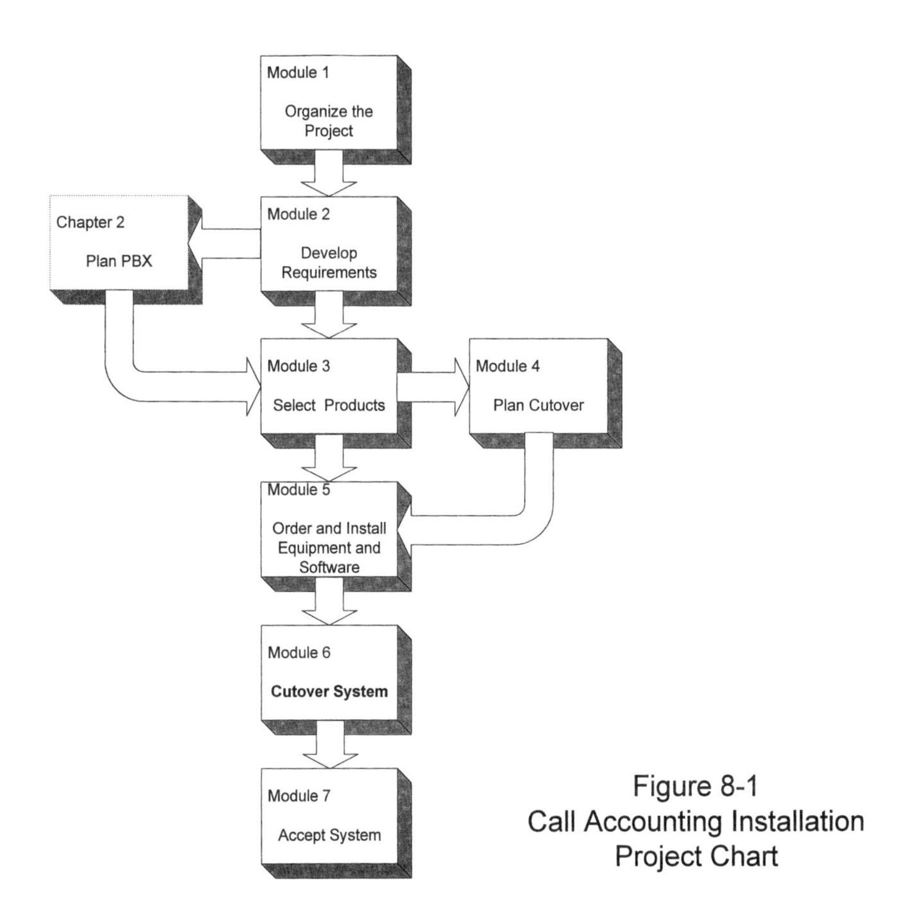

Figure 8-1
Call Accounting Installation
Project Chart

Module 1: Organize the Project

Set project objectives

Determine key objectives for the project. These must include completion date plus dates for any intermediate tasks that may affect target completion dates. Determine budgets for the project.

Organize project team

Identify all who are needed in the project. Representatives should be identified from the PBX vendor. Identify internal departments that must be represented. Typically, these include the telecommunications, information systems, facilities, and in some projects, human resources and public relations. Be certain that roles and responsibilities of all team members are clearly understood and accepted.

Hold kickoff meeting

In the initial project meeting all team members should understand their own and others' responsibilities. The kickoff meeting has the following objectives: assign and accept key responsibilities, communicate objectives and constraints of the project, establish schedules and content of reports and project meetings. Someone should be assigned responsibility for preparing minutes and distributing them within a day or two of the meeting.

Set cutover dates

Determine key dates in the project, including the final cutover dates as well as any interim dates that must be met.

Schedule project team meetings

Develop a schedule and place for future team meetings. The frequency of meetings depends on the competence of the project team, the complexity of the project, the penalties for missing due dates, and the numbers of people who must be kept informed.

Project Team Organization Complete

Identify constraints

Determine from all project team members any constraints that their company has with respect to force availability, inability to work certain dates, availability of key personnel or test equipment or other factors that may affect project completion. Document all such constraints in the minutes

Develop task list

Develop a detailed list of tasks that must be performed. Assign each task to a responsible individual. Sequence tasks in the order in which they must be performed. Obtain estimates of the amount of time required for each task.

Develop detailed cutover schedule

Identify milestones in the schedule. Set dates for completion of each milestone. Determine the critical path and determine whether the schedule fits within the objective interval. If it does not, determine how to compress the schedule until it fits. Create Gantt and/or PERT charts showing task sequence and schedules, and distribute to all team members.

Project Schedule Complete

Module 2: Develop Requirements

Determine technical requirements

Determine the computer and operating system the product must run on. Many products run on Microsoft Windows 95, which requires substantial processing power. Determine what types of calls will be recorded. Choices are a combination of long distance, total outgoing, total incoming, and in some systems, internal calls. Determine requirements for registering caller ID on the output reports. Determine the need for LAN access. Network capability may be required if several users need to access information in the database.

Determine telemanagement requirements

Determine what, if any, telemanagement features are needed in the product. User billing for a monthly rate in addition to toll costs is a common feature. Determine what features will be billed. Determine what process you want to use in getting billing changes into the system. This may lead to the need for a work order module. Evaluate the need for wire management, trouble reporting, and other telemanagement modules.

Determine system architecture

Call accounting systems come in a variety of configurations, ranging from a single PC connected directly to the PBX, to a fully networked system with a shared database on a file server, and a separate polling computer that downloads information from solid-state buffers attached to multiple PBXs. Determine which will be used.

Determine need for toll fraud detection

Many call accounting systems monitor calls in real time and flag the administrator in case of suspected toll fraud. Determine whether such is required, and what type of alerting is needed. Dial-out to pager is common.

Determine call rating requirements

Call accounting systems provide a variety of methods of rating calls. Vendor rate tables may be installed in the system and selected on installation. Rating tables may be based on V&H coordinates, which is an accurate method of computing airline distance between points. This method is unnecessary, however, if the company's carrier bills on flat rate exclusive of distance. Determine the need to bill for local calls.

Determine markup requirements

Some organizations mark up the cost of calls to help recover the cost of trunks, switching equipment, and administration. Hotels and similar companies plan to profit from marking up calls.

Determine need for reports on demand

Hotels, hospitals, dormitories, and other organizations with transient population may need to produce immediate reports for checkout. Immediate reports may also be needed for checking for toll fraud.

Determine tie line reconciliation requirements

When multiple PBXs are networked together, it is common for some PBXs in the network to complete their long distance calls over trunks attached to another PBX. In such a case, two records are produced for each call: one record from the originating PBX and the other from the terminating PBX. Some PBXs pass the originating station number across the network, but others do not. If both the originating station number and the terminating trunk group are required for rating the call, tie line reconciliation software may be required. Tie line software matches call records and creates a single record from them.

Requirements and Specifications Complete

Module 3: Select Products

Write a request for quotations or proposals

An RFP or RFQ is issued to identify suitable products specifying the requirements developed in the previous section.

Issue RFP

The RFP is issued to qualified vendors. Set a due date for responses.

Develop evaluation criteria

Determine how the product will be selected. Price is always a factor, but other factors should be considered such as ease of use, effectiveness of reports, and references from users with similar configurations.

RFP Responses Received

Evaluate responses

Vendor responses are reviewed. Non-conforming products are eliminated. The top two or three products are identified for further analysis.

Demonstrate products

The vendor demonstrates the product in an environment similar to the one in which the products will be used. Review the reports the system produces to be sure they are adequate. Try the major functions of the product to be sure that screens are logical, and that menus are easy to follow.

Check references

Discuss with at least three customers their experience with the product. Find out what experience they have had with the software developer's support. Get their opinion on ease-of-use in day-to-day operation. Ask whether they would buy the same product again. Determine whether they are using all telemanagement modules, and if not, discuss each module with someone who has used it.

Equipment Selected

Module 4: Plan Cutover

Select equipment space

Determine where the call accounting computer will be located and provide the necessary floor space and furniture.

Determine administrative methods

Determine who will administer the call accounting system. Determine how work orders for adding, moving, or changing stations will be entered into the call accounting system.

Determine database origin

Determine where the master list of stations and departments will be obtained from. The best source is download from the PBX if possible. Departmental assignments must come from another source.

Determine reports

Call accounting and telemanagement systems are capable of producing far more reports than most organizations need or will use. Review the reports produced and determine which ones will be regularly printed and how they will be distributed.

Determine call rating methods

Determine whether V&H or flat rating will be used. Obtain rate tables from the IXC and LEC. Determine whether calls will be marked up.

Plan transition from old system

If an existing call accounting system is in place, plan the physical move. Consider how to handle calls in progress if the company operates 24 hours per day. Determine how reports from the old system will be merged with reports from the new. Plan how to convert telemanagement database from the old system to the new. If the same computer will be used for the new system, plan for the software load and database transfer.

Cutover Planning Complete

Module 5: Order and Install Equipment and Software

Order the call accounting system

The call accounting/telemanagement product is ordered from the vendor, who orders it from the manufacturer.

Order computer equipment

Order any additional equipment needed for implementation. This may include a PC and printer, and designation of the necessary work space. Get the necessary cables and determine how the PC will connect to the PBX. Drivers may be required.

Equipment Ordered

Deliver call accounting system

The call accounting software is shipped from the developer and delivered to the site.

Deliver computer equipment

The PC and printer plus any ancillary equipment are purchased. Unbox the equipment and install in the space provided.

Equipment On Site

Power up the PC

Connect the PC to AC power or UPS, turn it on, and run through the boot-up process. Load the operating system software.

Load system software

The call accounting/telemanagement program is installed in the PC. The system is configured, and a record of the configuration is printed and retained.

Prepare user database

A call accounting record is established for each user. The most effective source of the record is the PBX database, so if possible, the call accounting database is created from a downloaded copy of the PBX database. The organization structure must also be loaded to enable the system to create reports by organizational unit.

Prepare trunk database

The system requires a database of all trunks connected to all PBXs in the network. The type of trunk is needed, and the group of which each trunk is a member is required. The trunk group is named for ease of identification.

Prepare telemanagement database

If telemanagement features are used, additional database conversions will be required. For example, a database of all cable pairs may be required to support the cable assignment process. A cable and pair numbering system may have to be created if it doesn't already exist. Other features such as work orders and trouble reports may operate off the database used for call accounting.

Prepare call rating database

Rates are loaded into the database in accordance with the rates your carrier charges. Load the percent discount or markup against the carrier's charges.

Review databases

Copies of the databases are printed and reviewed for accuracy. The trunk and rating databases are the most important because they affect billing. Where calls are being charged to users as, for example, in hotels, hospitals, universities, and shared tenant services, this review is of the utmost importance because it affects the amount collected. In other commercial organizations, the primary effect of database errors is inaccurate distribution of costs to departments.

Load call accounting databases

The user, trunk, organization, call rating, and telemanagement databases are loaded into the call accounting PC. Review all screens to check accuracy of the load.

Connect the system to the PBX

If several people will use the system, connect it to a LAN. Connect call storage buffers if these are used, and connect them to analog ports on the PBX. Assign DID numbers to the buffers. If necessary, connect a remote access modem to the PC so it can be supported remotely.

Program the PBX for call detail recording

The PBX is programmed to send the appropriate call detail (long distance, local, etc.) to the CDR port. Check the call storage buffer or PC connected to the CDR port to be certain data is flowing.

Configure polling PC

If call details are being polled from storage buffers or remote PCs, configure the polling PC. The polling device must be connected to an outgoing telephone line, programmed with the number or numbers to be polled, and programmed as to the time of day polling is to occur. Balance the frequency of polling against the possibility of data loss. The polling frequency must be set frequently enough to ensure that buffers do not overflow.

Test polling operation	If call details are being polled from a call storage buffer or remote PC, test the polling PC to be certain it is downloading data.
Load client software	In telemanagement systems, load the application software on the users' PCs. Configure and test for proper operation.
Test call accounting system	The system is tested to verify that it polls properly, processes and rates calls accurately, stores data to the hard disk, and correctly processes telemanagement functions such as service orders, cable assignments, and trouble reporting.

Module 6: Cutover System

Train users of telemanagement system	Deliver user training. Hand out instructional material if provided. Hand out client software and instruct users how to install on their systems and use it.
Cutover system	The server is placed in service and tested thoroughly for all functions.

Cutover Complete

Module 7: Accept System

Evaluate service

Conduct post-cutover service evaluation. Determine whether the system is meeting its objectives. Check accuracy of call rating. Check toll fraud alerting equipment. Test reports on demand.

Accept system

When the system has met all of its requirements, it is accepted, and final bills are paid.

Project Complete

Chapter 9
Local Area Networks

"...There is nothing more difficult to plan, more doubtful of success, nor more dangerous to manage than the creation of a new system. For the initiator has the enmity of all who would profit by the preservation of the old institution and merely lukewarm defenders in those who would gain by the new one."

Machiavelli, *The Prince*, 1513

LANs have become as vital for data communications in the office as PBXs are for voice communications. As LAN operating systems become more complex, the telecommunications manager is likely to experience several installations and upgrades in the course of a career. The technology is advancing so rapidly that the lessons learned from one installation are almost lost on the next. Different versions of network operating systems bear little resemblance to one another, and the LAN hardware evolves rapidly as network cards conform to plug-and-play standards. The span of the LAN also is evolving as companies use the LAN internet to connect offices together.

The LAN project follows a structure similar to other telecommunications projects. Modules cover project team formation, requirements development, and equipment acquisition. Equipment installation involves file servers and workstations as separate entities. The installation includes printers and other peripherals. The project may also coordinate with or include details from wiring, and backbone cabling, which are included in other chapters. Most LANs in large companies will also involve an internet, which is covered in Chapter 11.

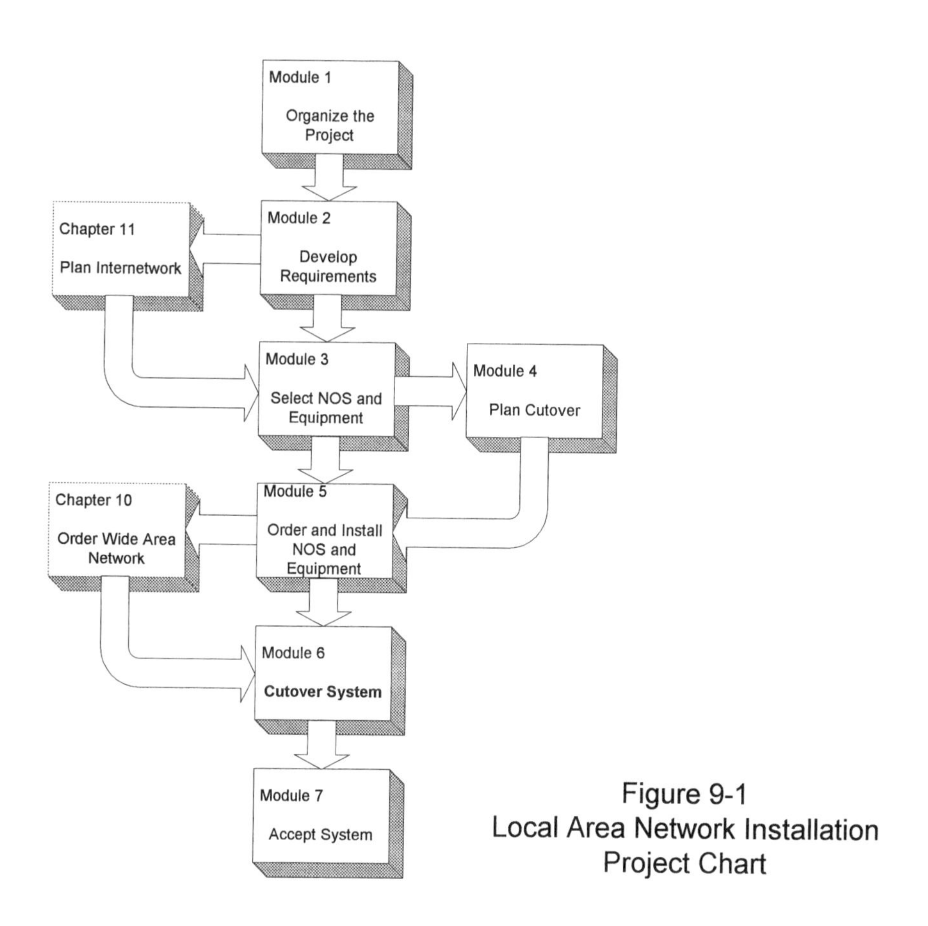

Figure 9-1
Local Area Network Installation
Project Chart

Module 1: Organize the Project

Set project objectives

Determine key objectives for the project. These must include completion date plus dates for any intermediate tasks that may affect target completion dates. Determine budgets for the project. Determine the span of the network. Is it local, or will it interconnect with LANs in branch offices?

Organize project team

Identify all who are needed in the project. If rewiring is involved, the wiring contractor should provide a representative. If an outside vendor will install the network equipment and operating system, a representative from the vendor is required. Identify internal departments that must be represented. Typically, these include the telecommunications, information systems, facilities, and in some projects, human resources and public relations. Be certain that roles and responsibilities of all team members are clearly understood and accepted.

Hold kickoff meeting

In the initial project meeting all team members should understand their own and others' responsibilities. The kickoff meeting has the following objectives: assign and accept key responsibilities, communicate objectives and constraints of the project, establish schedules and content of reports and project meetings. Someone should be assigned responsibility for preparing minutes and distributing them within a day or two of the meeting.

Set cutover dates Determine key dates in the project, including the final cutover dates as well as any interim dates that must be met.

Schedule project team meetings Develop a schedule and place for future team meetings. The frequency of meetings depends on the competence of the project team, the complexity of the project, the penalties for missing due dates, and the numbers of people who must be kept informed.

Project Team Organization Complete

Identify constraints Determine from all project team members any constraints with respect to force availability, inability to work certain dates, availability of key personnel or test equipment or other factors that may affect project completion. Document all such constraints in the minutes

Develop task list Develop a detailed list of tasks that must be performed. Assign each task to a responsible individual. Sequence tasks in the order in which they must be performed. Obtain estimates of the amount of time required for each task.

Develop detailed cutover schedule Identify milestones in the schedule. Set dates for completion of each milestone. Determine the critical path and determine whether the schedule fits within the objective interval. If it does not, determine how to compress the schedule until it fits. Create Gantt and/or PERT charts showing task sequence and schedules, and distribute to all team members.

Project Schedule Complete

Module 2: Develop Requirements

Determine applications that will run over the network

LAN architecture depends on applications. Database applications, in particular, have the potential of overloading the network unless it is segmented or connected through switches. Ordinary office applications such as spreadsheets and word processing may require large amounts of file storage space, but they put a light load on the network. E-mail is an important application of most networks.

Determine workstation requirements

Determine the workstation operating software and the applications that each work station will support. Determine the hardware platform of the workstations. Determine processor speed, amount of RAM, amount of disk storage space, type of monitor, and other variables pertaining to the workstation.

Determine workstation locations

Workstations must be physically connected within the wiring limits of the transmission medium. For twisted-pair wire, the limits are generally 90 wire meters from workstation to hub. When greater distances are required, hubs are linked with fiber optics or with routers or bridges over public network facilities.

Determine network transmission medium

Choices of transmission medium are shielded or unshielded twisted-pair wire, thin and thick coaxial cable, fiber optics, and wireless. The choice is dictated by building structure, compatibility with existing equipment, range requirements, and requirements established by choice of network protocol.

Determine bandwidth requirements

Light and medium duty LANs usually operate at 10 mb/s for Ethernet or 4 or 16 mb/s for token ring. Networks requiring greater bandwidth can use a variety of other protocols such as 100 mb/s Ethernet, 100 VG, FDDI, or ATM. The bandwidth requirements are driven by immediate needs and projected growth.

Determine LAN architecture

The LAN architecture can range from a single segment to multiple segments in the same building connected by routers, switches, or both. Off-site LANs can be connected over public networks. The objective is to design the total network with all major elements in place.

Determine subnet architecture

When LANs exceed the limits of a single subnet because of physical wiring limits, network load, or both, the network is segmented into subnets. The number and size of segments affects the need for routers, switches, or bridges.

Determine switching requirements

Determine the number of ports that must terminate on switches and whether switches must support standard or fast Ethernet or token ring. Determine whether switches should be cut-through or store-and-forward. Determine switch management requirements; i.e. should switches be RMON equipped.

Determine hub configuration	LANs operating over twisted-pair wire require centralized hubs or switches. Determine whether hubs, switches, or both will be used.
Determine hubbing hierarchy	In buildings with multiple hub locations, hubs can be hierarchically arranged (i.e. ports connected so one hub operates as a slave from another. Be careful not to violate Ethernet segment rules).
Determine style of hubs	Hubs are furnished in several styles, most important of which are modular wiring concentrators and stackable. Modular hubs have card slots into which different types of cards such as Ethernet, fast Ethernet, Token Ring, and AppleTalk can be plugged. Modular hubs have multiple backplanes for ease in segmenting networks. Stackable hubs contain a fixed number of ports of uniform type. They are expanded by adding hubs of the same type.
Provide space for hubs	Provide relay rack or backboard space for mounting hubs. Space must be within patch cord reach.
Determine printing requirements	At least one shared printer is needed in each LAN segment. Determine the type and speed of the printer, paper capacity, special fonts, and other printer capabilities. Determine how printers will connect to the network. In most cases connections will be to a print server, but in some networks shared printers

will attach to a workstation. Determine requirements for other related devices such as plotters and color printers.

Determine expected network load

The load on a network segment is affected by the number of workstations, the applications they are running, and the amount of activity from each station. It is also affected by the extent to which files and software are stored on local hard disks and how much are kept on file servers. The load should be expressed in bits or packets per second during peak intervals and compared to the effective throughput of the network protocol. If the load exceeds the capacity, further segmentation is required.

Determine workstation software standards

Establish standards for applications that will be supported on network workstations. Determine whether applications will be stored in the server or on the workstation hard disks.

Choose the workstation operating environment

Workstations can run under DOS, one of the Microsoft Windows versions, or under a Macintosh operating system. The choice of operating system is a function of the choice of application software and the resources of the workstation. The workstation operating environment may provide peer-to-peer capability, which permits workstations to offer resources such as files and printers for sharing.

Select network operating system

The network operating system is the heart of the LAN. It is chosen based on criteria such as the availability of support, ease of administration, security provisions, availability of network utilities such as fault-locating and service evaluation tools, and the ability to interoperate with other networks in the organization's environment .

Determine the type of network operating system

The two major types of NOS are server-centric and peer-to-peer. Peer networks are suitable for small offices with minimal security requirements. Larger offices and those where files must be secured will choose a server-centric NOS.

Determine file server requirements

Determine whether single or multiple file servers are required. For each server, determine the amount of RAM required, the size of the hard disk, and type of interface. Interface types are IDE or SCSI. Determine bus requirements on the server (EISA ISA, PCI, etc.). Determine the need for peripherals such as tape backup and CD/ROM. Determine whether disk duplexing or mirroring is required. Determine the speed and type of the network interface cards required (fast Ethernet, full-duplex Ethernet, iso-Ethernet, token Ring). Determine the type and speed of the processor and whether multiple-processors are required. Determine monitor requirements (type, size, dot density, etc.).

Determine specialized server requirements

Determine the need for specialized devices such as fax servers, modem pool, and CD/ROM servers. Determine where these will be located and what segments they will connect to.

Determine internetworking requirements

Determine whether the LAN must be linked to other networks such as LANs at other company locations or to the Internet. If internetworking is required, determine the traffic volume between networks and use this to determine the bandwidth of circuits required. Determine whether public facilities will be used or whether private facilities are available.

Determine need for mainframe connectivity	Many local area networks are developed for the purpose of enabling multiple users to connect to mainframe and minicomputers. Determine whether mainframe connectivity is required, and if so, what communication protocols the mainframe requires. Determine whether connectivity can be supported directly by connecting hardware devices to the LAN or whether a gateway is required.
Determine backbone architecture	If internetworking is required, determine the internetwork backbone architecture. Choices are distributed or collapsed backbone, switching, or a combination.
Determine backbone protocol	For distributed backbone networks, determine the protocol. (for example, FDDI, ATM, etc.) that will be used.
Determine domain name and IP addressing structure	If the network will be connected to the Internet with this project or in the future, and if TCP/IP protocol will be used, obtained an Internet domain name and block of IP addresses.
Develop IP addressing system	Develop IP addressing plan, either as part of an existing numbering system, or initiated with this project.
Determine station wiring requirements	Determine whether the existing wiring is adequate to support the LAN. If not, a plan for upgrading the wire is required. See Chapter 13 for additional information on wiring systems.

Determine network security requirements

Determine the need for hidden directories and for password access to visible files. Determine password length and frequency of forced password change. Determine the need for firewall security for connections to the Internet or remote access servers.

Determine type of public network facilities

If public network connections are used for LAN internets, determine what type of facilities will be used. The primary choices are dedicated digital facilities (T-1, fractional T-1, or 64 kb/s), frame relay, and ISDN. If frame relay is used, determine the CIR, port speed, and access circuit bandwidth. See Chapter 10 for further details on frame relay.

Determine remote access requirements

Determine whether off-site access to the network is required for functions such as telecommuting, remote customer access, off-site Internet access, or other such applications. If remote access is required, determine whether dial-up access will be analog or ISDN. See Chapter 7 for further discussion on remote access.

Determine administrative strategy

Determine who will be responsible for ongoing administration of the network. Duties include adding and removing users, administering security, backing up servers, detecting and resolving users' problems, setting up workstation software, and other details of running the network. Determine how the network administrator will be trained and the degree to which he or she will be involved in the initial configuration.

Determine outside support requirements

Determine the amount and type of vendor support that you will require for the network. Most companies require installation support. In addition, many companies contract for administrative support. Determine whether support must be provided by the vendor that provides the network equipment, or if a third-party source is available.

Determine network management requirements

Most network elements can be managed remotely over they network if the elements are equipped for management using SNMP. Network statistics can be collected from managed elements if they are compatible with RMON. Determine which network elements need to be managed and whether they must be RMON-compatible. Determine quantities and locations of network management workstations. Determine whether management workstations will be UNIX workstations or PCs.

Determine NMS workstation requirements

For the NMS workstation, determine the amount of memory, hard disk space, screen size, processor speed, and other such variables.

Determine UPS requirements

Determine the location, capacity, and power conditioning requirements for the file servers' UPS. Determine the need for SNMP management and for automatic server shut down.

Develop a directory structure

Develop a directory hierarchy for file servers. Develop naming conventions for the directories and files.

Develop a security plan for the file server

Determine the rights and permissions that each user has to elements of the directory structure. Determine directories that are to be hidden from users without permission.

Determine which directories or files will be classified as read-only. Determine the password structure for the network. Establish supervisory passwords. Determine if departmental supervisors will be used, and establish the limits of their authority. Set policies on password length, composition, and frequency of change.

Requirements and Specifications Complete

Module 3: Select NOS and Equipment

Prepare request for proposals

Prepare a request for proposals or other procurement document that expresses all the requirements.

Issue RFP

Issue the RFP to qualified vendors. Set a due date for responses.

Identify product selection criteria

Criteria for choosing the winning proposal are identified. Consider such items as quality of vendor support, references, cost, provisions of desirable and value-added features, etc.

Receive LAN proposals

Vendors submit proposals for the local area network.

RFP Responses Received

Review LAN proposals

Proposals are opened and checked for conformance to the requirements and specifications. Nonconforming proposals are eliminated. Conforming proposals are compared to the evaluation criteria and ranked. The top two or three proposals are selected for further evaluation.

Select vendor

From the proposals and, if appropriate, vendor demonstrations and reference checks, select the vendor that will provide the network equipment and software.

Equipment Selected

Module 4: Plan Cutover

Develop network cutover plan

If the project is a new network where none has existed before, the cutover can be a simple matter of putting the network in service in a single "flash" cut. If an internet is being established to remote networks, more planning is required. The cutover team must decide whether to bring remote locations on gradually or as part of the initial cut. Where the network is transitioning from an existing system to a new and perhaps larger one, step-by-step procedures and plans will be required. The sequence of operations such as adding WAN circuits, placing and configuring routers, connecting mainframe computer links, and the like may be critical if all workstations are to be operational without any service interruption. The cutover plan should be a series of steps that all participants agree will achieve the objective with a minimal impact on service.

Develop network testing plan

Develop a plan for testing all functions of the network before it is placed in service. Each workstation should be tested for the ability to log on to each authorized server. Routers should the tested for connectivity to remote sites. All printers should be tested for accessibility from authorized locations. Security provisions should be tested. The objective of the test plan is to minimize the amount of lost time in instructing users and the need for on-site user assistance for the first few days following cutover.

Develop backup strategy

Determine how the file servers will be backed up. Determine what type of hardware will the used (tape, DAT, removable disk, etc.). Determine how backed-up files will be removed from the building and stored off-site.

Determine file server location

Determine where the file server will be installed. The location requires physical space for both the servers and UPS, a dedicated electrical circuit with isolated ground, adequate ventilation, cooling if necessary, and adequate physical security.

Develop training plans

If users on the network are not familiar with operating on a LAN, training should be provided. Develop training plans for instructing users on concepts of resource redirection, log on and log off, directories, and the disciplines of functioning in a network environment. Consider training requirements for network administrators.

Cutover Planning Complete

Module 5: Order and Install NOS and Equipment

Order network hardware

The customer orders hubs, routers, switches, NICs, etc. from the vendor. Vendor orders from manufacturer.

Order network operating system

The customer orders the network operating system from the vendor, who orders it from the manufacturer.

Order server

The customer orders the server from the vendor who assembles it or orders it from the manufacturer.

Equipment Ordered

Order public network circuits

If the network is interconnected with other LANs over public facilities, place network orders with the service provider. See Chapter 10 for additional details on wide area network installation.

Install wiring

Install wiring to the workstations. Terminate on patch panels in the equipment room and telecommunications closest. Perform Category 5 tests on wire. See Chapter 13 for additional details.

Install backbone cabling

Install fiber optic or other cabling used for connecting hubs, switches, and routers to the backbone. Test backbone in accordance with testing plans. See Chapter 11 for additional details.

Deliver network hardware	The vendor delivers NICs, hubs, routers, switches, etc.
Deliver network operating system	The vendor delivers the network operating system.
Deliver server	The vendor delivers the server.

Equipment On Site

Install network hardware	Install hubs, routers, switches, and other network hardware. Connect hubs and switches to workstations with patch cords.
Install server	Server is unboxed and any necessary peripherals are installed. The network interface card is added. The server is connected to AC power through the UPS and powered up.
Install network operating system	The NOS is installed in the server in accordance with the manufacturer's instructions. The NOS is configured to meet the customer's requirements. Set up system log-on scripts
Set up directory structure	The customer's desired directory structure is programmed in the file servers.
Install security on servers	Rights and permissions are programmed into the directory structure. Supervisor's ID and password are programmed. Trustees' permissions are programmed.
Establish computer names	Computers use MAC or IP addresses to identify each other, but users work with computer names. Establish domain names, group names, server names, and workstation names. In large networks the workstation

	name is usually a contraction of first and last names.
Create users in servers	Users are identified and programmed into the operating system. Users groups are defined in the NOS, and users are added to user groups.
Create user environment	Develop the log-on scripts, icons, virtual drives, and other aids to make it easy for the users to operate on the network.
Determine printer locations	Locate printers to provide maximum convenience for users. Choose locations that are convenient for those who tend the printers and deliver output.
Install hubs and switches	Install hubs and switches in the relay racks or shelf space provided. Connect to power and connect to patch panels with station cables.
Connect network printers	Connect network printers to the station cabling in the case of direct connected printers or to a print server or file server as appropriate.
Configure network printers	Set up printer name and IP address and configure the printer for such variables as banner pages, paper size, etc. If other peripherals such as plotters are used, set them up accordingly. Set up print queues and designate certain users as printer administrators, which gives them the ability to delete jobs from queue and reprioritize print jobs.
Set up workstations	For existing workstations, install NICs. For new workstations, unbox computers, install

NICs, and install operating system software. For both new and old computers, set up workstation software including log-on scripts, printer redirection, virtual drives, work station names, IP address, and other variables to make the work station operational. Connect the workstation to station cabling and verify that the station has connectivity to the hub or switch. Log on all servers for which the workstation has permissions. Shut the workstation down to verify the shutdown and log-off process. Print a test page to each network printer for which the workstation is configured.

Set up internetworking equipment	If the network is interconnected with routers, switches, or bridges, set them up according to the manufacturer's instructions. See Chapter 11 for additional details.
Set up software metering.	Many networks use software metering to comply with licensing requirements without purchasing individual packages for each user. Software metering allows the programs to be stored in file servers with the number of licensed copies identified.
Install network management system	Install NIC in the NMS computer. Connect to power and load management software. Connect the computer to the network.
Configure network management system	Run manufacturer's boot-up tests on the NMS. Assign the IP address. Configure the system according to manufacturer's instructions.
Test network operations	Before the network is turned up for service, all operations of the network are tested in

accordance with the testing plan. Any deviations are corrected.

Installation Complete

Module 6: Cutover System

Train users

Train users on how to log on and off of the network, use directories, shared resources, and network security. Cover items such as password creation and changes, personal file space, loading software from the network, and other operational functions.

Transfer files to the file server

If an existing server is being replaced, transfer the files from the old server to the new directory structure. Take the opportunity to purge outdated files. Back up the old server before files are transferred.

Set up help desk

For users who are unfamiliar with network operations, a special help desk should be set up. When the help request load drops to a reasonable level, the help desk reverts to normal operations.

Cutover the network

Following the cutover plan, the network is placed in service.

Cutover Complete

Module 7: Accept System

Evaluate service

When the network has been in operation for a few days, a formal evaluation of service should be made. Contact users to determine if the network is meeting their expectations. Evaluate reports from the help desk to determine whether an acceptable level of trouble exists. Review records of hardware faults to detect excessive failures. Determine whether the vendor has met its commitments. Continue monitoring the network until acceptance criteria has been fulfilled.

Accept system

After all criteria have been fulfilled, the system is accepted and final payments are made.

Project Complete

Chapter 10
Wide Area Networks

"It is a bad plan that admits of no modification"

Publius Syrus Maxim 469

Wide area networks carry data, voice, or a combination of both. Sharing between voice and data is common in the access circuits, and less common in the interexchange circuits, where switched voice traffic is inexpensive. Implementing both voice and data circuits involves many steps that must be carefully coordinated. It is particularly important to ensure that circuits and equipment arrive at the same time to prevent expensive resources from lying idle while waiting for a critical element to show up.

To design a network, the designer collects information about the amount of traffic flowing between locations. With this information the designer determines the architecture of the network and what kind of equipment it requires. Data network architectures are changing significantly with the advent of new services and the replacement of dumb terminals by personal computers. In data networks, circuit costs and equipment configurations are changing to favor frame relay, although some networks continue to use dedicated point-to-point circuits or X.25.

With voice WANs the service terminates in a PBX, but the architectures are likewise changing. In the past, PBX networks were interconnected with analog or T-1 tie lines. Switched service costs are dropping to the point that only the largest companies can justify the cost of switched voice circuits.

Both data and voice network design involves obtaining circuit quotations from carriers. Circuit costs are no longer distance sensitive to the degree they once were. Also, tariffs are disappearing for many services, with commercial contracts taking their place. Therefore, the only way to obtain reliable cost data on any kind of voice or data circuits is to obtain quotes directly from the carriers. Many of the project management techniques are the same for both voice and data networks. This chapter separates data WANs from voice to identify those factors that are different.

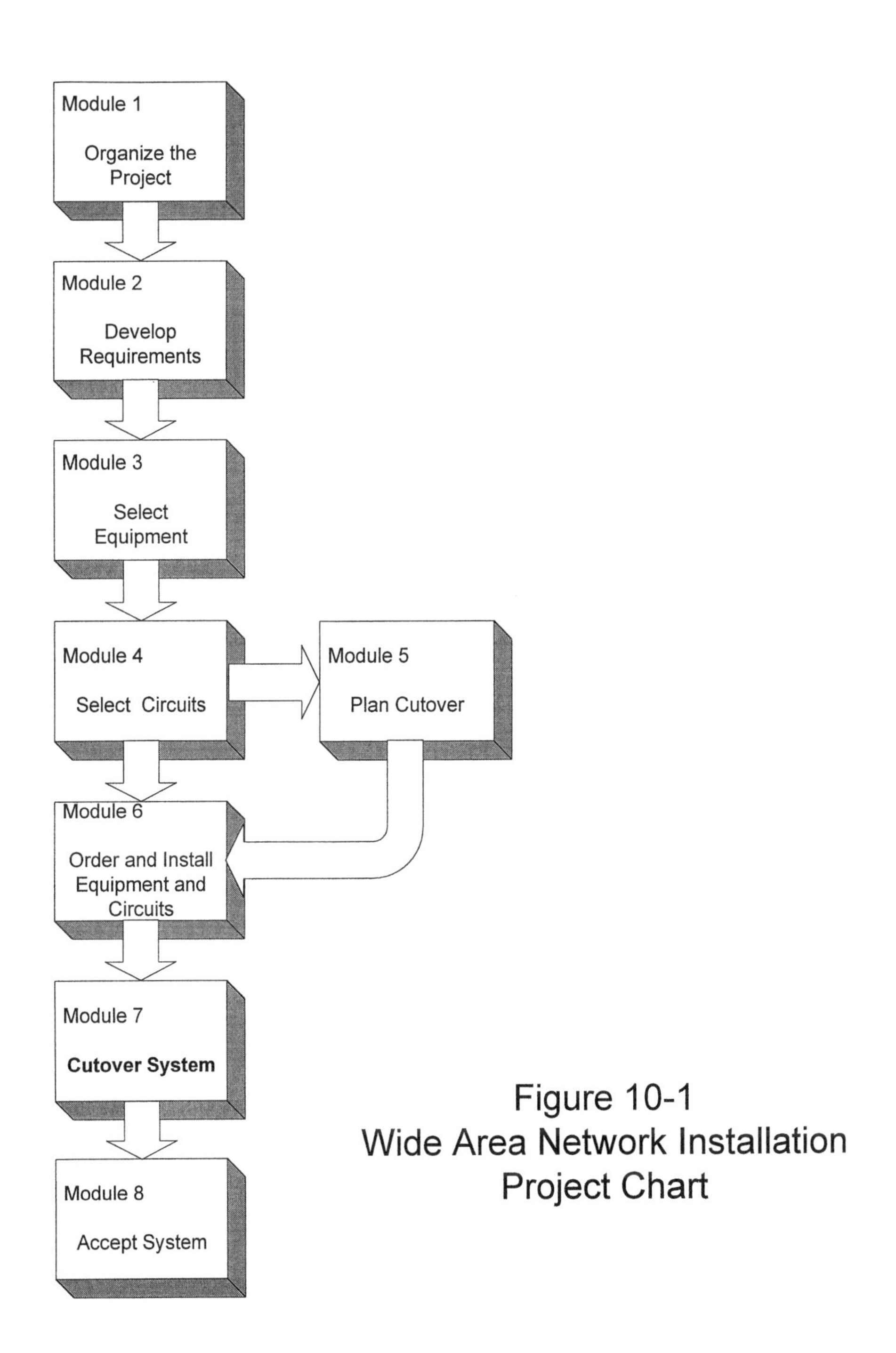

Figure 10-1
Wide Area Network Installation Project Chart

Module 1: Organize the Project

Set project objectives

Determine key objectives for the project. These include completion date plus dates for any intermediate tasks that may affect target completion dates. Determine budgets for the project.

Organize project team

Identify all who are needed in the project. Representatives should be identified from the LAN and network equipment vendors plus representatives from the carrier that will provide circuits. Identify internal departments that must be represented. Typically, these include the telecommunications, information systems, and facilities departments. Be certain that roles and responsibilities of all team members are clearly understood and accepted.

Hold kickoff meeting

In the initial project meeting all team members should understand their own and others' responsibilities. The kickoff meeting has the following objectives: assign and accept key responsibilities, and communicate objectives and constraints of the project, establish schedules and content of reports and project meetings. Someone should be assigned responsibility for preparing minutes and distributing them within a day or two of the meeting.

Set cutover dates

Determine key dates in the project, including the final cutover dates as well as any interim dates that must be met.

Schedule project team meetings

Develop a schedule and place for future team meetings. The frequency of meetings depends on the competence of the project team, the complexity of the project, the penalties for missing due dates, and the numbers of people who must be kept informed

Project Team Organization Complete

Determine objectives

Determine the service objectives the network is required to meet. Computer network service is measured in terms of response time, percent of network availability, and minimum throughput requirements. Determine the required in-service date. Determine growth expectations and the expected life of the service for which the network is being developed.

Identify constraints

Determine what constraints affect the network design. Examples are protocols that must be employed or avoided, existing equipment that must be reused, peak work load times during which no disruption is permitted, etc.

Develop task list

Develop a detailed list of tasks that must be performed. Assign each task to a responsible individual. Sequence tasks in the order in which they must be performed. Obtain estimates of the amount of time required for each task.

Develop detailed cutover schedule

Identify milestones in the schedule. Set dates for completion of each milestone. Compute the critical path and determine whether the schedule fits within the objective interval. If it does not, determine how to compress the schedule until it fits. Create Gantt and/or PERT charts showing task sequence and schedules, and distribute to all team members.

Project Schedule Complete

Module 2: Develop Requirements

Determine network service locations

Determine the locations of all points on the network that must be connected. Obtain the NPA and prefix of the serving central offices (required for obtaining the costs of service).

Determine bandwidth requirements

Determine the amount of traffic flowing during peak periods between all locations. Determine which locations must communicate with which other locations, and the amount of data flow between each pair of stations. Data flow should be estimated in peak hour bits per second. Allow additional capacity for growth. Allow additional bandwidth for bursty traffic.

Determine protocol

Determine which protocols the network

requirements must support. If the company is using obsolescent protocols that will be replaced, determine how long such protocols must be kept operational.

Determine architecture of existing network Determine what current networks are in place, and which of these must be retained. Obtain circuit designations, equipment nomenclature, applications, and other factors that could have an impact on the design.

Determine voice/data sharing objectives Determine the degree to which the network can be shared between voice and data. T-1 lines to the IXC can often be split between voice and data to reduce access circuit costs.

Determine security requirements Where networks are connected to public networks through remote access servers (see Chapter 7) or to the Internet, security is of the utmost importance. Determine the need for firewalls, which are routers or specialized devices that filter packets to prevent outsiders from accessing restricted portions of the network.

Determine node locations Many data network architectures identify hubbing locations and run tail circuits to other locations. Identify the hubbing locations, which are usually chosen for their effect on circuits costs and because they represent major concentrations of usage.

Determine facility types Determine what types of facilities will be used for the service. The most common services are dedicated digital circuits (56 kb/s, T-1, T-3) and frame relay. Analog circuits may be required for additions to existing networks, and high-bandwidth services such as ATM and SMDS may be appropriate.

Verify circuit availability

Verify that the circuit types assumed in the design are available in all locations.

Determine network management requirements

Determine how the network will be managed. Determine which devices are to be SNMP compatible. Determine the need for RMON. Determine locations and quantities of management workstations. Determine where alarms will terminate. Determine how trouble will be isolated and dispatched. Determine help desk requirements.

Develop disaster prevention requirements

Determine what steps are needed to prevent loss of service during unusual conditions. Determine the need for self-healing access circuits, services to alternate carriers' points-of-presence, and other such features that carriers offer to prevent loss of service.

Develop configuration information

Determine information required to configure circuits and equipment. Determine IP addresses, DLCI, T-1 framing type (D-4 or ESF), etc.

Requirements and Specifications Complete

Module 3: Select Equipment

Determine equipment requirements

Determine types and quantities of equipment such as bridges, routers, CSUs, hubs, etc. Determine the specifications such equipment must meet such as the minimum packet forwarding rate, protocols that must be supported, and the number and type of circuits each device must interface.

Evaluate cost/performance characteristics of alternative designs

Review different design alternatives to determine which produces the best combination of price and performance. Detail the differences between alternatives and obtain approval to proceed.

Write equipment request for proposals

Prepare a request for bids or proposals to provide network equipment. Establish a date for required responses. If appropriate, include a requirement for vendor installation and setup. Issue the RFP to selected vendors. In some cases both circuits and equipment can be provided by the same companies.

Equipment RFP Responses Received

Review responses

Review vendor responses to the RFP. Reject nonconforming proposals. Rank the remaining proposals in order based on price, references, implementation capability, and other such criteria. Check vendors' references. Invite vendors to make product presentations if appropriate.

Select equipment vendor

Select the winning equipment proposal. Negotiate contract terms and conditions with the vendor.

Equipment Selected

Module 4: Select Circuits

Estimate costs

Determine circuit costs from carriers. Obtain an estimate of equipment costs from representative vendors.

Prepare circuit RFP

Some circuit types are covered by tariff and the cost is not negotiable. Increasingly, however, circuit costs can be negotiated, particularly when they are associated with a larger contract for long distance. Frame relay costs are almost always negotiable. In most cases the carriers offer discounts for term and volume commitments. Prepare a request for proposals or quotations and issue it to qualified carriers. Set a response deadline.

Circuit RFP Responses Received

Review circuit proposals

Open the proposals. Discard the proposals that fail to meet the specifications. Rank the remaining proposals by value. Pay close attention to the carriers' network architectures and assess them for reliability. Ring networks are generally the most reliable with spurs off the ring offering potential failure points.

Select circuit provider	Select the carrier that will provide circuits. Negotiate a contract if appropriate.

Circuits Selected

Module 5: Plan Cutover

Plan the project	Meet with the carrier, the equipment vendor, and others who have a part in the project. Develop a detailed plan for the remainder of the project including due dates, responsibilities, and reporting requirements.
Determine specific equipment configuration	Determine the quantities of circuit equipment such as bridges, routers, switches, CSUs, etc. that will be required to implement the system.
Determine specific network configuration	Develop a final configuration concerning major network node locations, circuit bandwidths, required due dates, etc.
Determine protocol configuration	Based on the circuit and equipment selected, determine the final protocols that must be supported. Develop IP addresses, determine DLCI if appropriate, plan routing tables, and specify protocol options.
Develop testing plan	Develop a plan for testing the network. Develop plans for testing security arrangements and for testing the network's capability of carrying the required load.

Develop synchronization plan

All devices on a digital network must be synchronized. Devices such as PBXs and routers take their timing from the source closest to the national reference frequency. Other devices on the network slave from the master. Be careful to avoid synchronization loops. Prepare a synchronization plan. Consult an expert if in doubt about plan validity.

Develop final security plan

Based on circuits and equipment selected, plan how security will be administered.

Develop disaster prevention plan

Develop a plan integrating equipment and circuit alternatives to assure service continuity through disaster conditions.

Cutover Planning Complete

Module 6: Order and Install Equipment and Circuits

Order equipment

Order network equipment from the selected vendor. Establish due dates to conform to the required service date and the date of circuit availability.

Order circuits

Order circuits from the selected carrier. Set due dates and obtain carrier concurrence that dates can be met.

Circuits and Equipment Ordered

Install equipment

Vendor installs network equipment in all locations. Install equipment in the sequence in which circuits will be turned up. See Chapter 11 for additional details on internetworking equipment.

Install circuits

Carriers install circuits in all locations. Carriers test circuits end-to-end for point-to-point circuits or from customer locations to the network node in case of frame relay or X.25 circuits.

Configure network equipment

Routers, switches, bridges, and other network equipment is configured in accordance with the network plan. Routing tables are built, IP addresses and DLCI are assigned, and other variables are programmed into equipment. Equipment is connected to circuits.

Equipment Configuration Complete

Test network continuity

Network connections are tested through the equipment end-to-end. Test each pair of termination points. Monitor circuits for bit error rate performance.

Stress test network

Send test data at speeds high enough to validate that the network meets design specifications. Refer any malperforming circuits to vendors to correct.

Set up firewalls

Set up the firewall device or configure firewall router to filter out non-permitted

	transactions. Test firewall to ensure that it is impregnable.
Set up user access	Configure all servers with the user rights and permissions from remote termination points. Assign passwords and log-ins. Reconfigure security on LANs.
Module 7: Cutover System	
Train users	Train users on how to log on to remote servers and hosts. Train users on network procedures and security.
Place the network in service	The network is turned up for service. Monitor the system closely for the first few days to be sure it is meeting expectations.

Cutover Complete

Module 8: Accept System	
Evaluate service	Conduct post-cutover service evaluation. Determine whether the system is meeting its objectives. Review statistics from the network to determine whether its usage is meeting expectations. If possible, interview users to determine their satisfaction with the system.
Accept network	The customer accepts the network from vendors, makes final payment on equipment, and places the network in service.

Project Complete

Module 1: Organize the Project

Set project objectives

Determine key objectives for the project. These include completion date plus dates for any intermediate tasks that may affect target completion dates. Determine budgets for the project.

Organize project team

Identify all who are needed in the project. Representatives should be identified from the switching system vendor and the IXC. Identify internal departments that must be represented. Typically, these include the customer service, telecommunications, information systems, and facilities departments. Be certain that roles and responsibilities of all team members are clearly understood and accepted.

Hold kickoff meeting

In the initial project meeting all team members should understand their own and others' responsibilities. The kickoff meeting has the following objectives: assign and accept key responsibilities, and communicate objectives and constraints of the project, establish schedules and content of reports and project meetings. Someone should be assigned responsibility for preparing minutes and distributing them within a day or two of the meeting.

Set cutover dates

Determine key dates in the project, including the final cutover dates as well as any interim dates that must be met.

Schedule project team meetings

Develop a schedule and place for future team meetings. The frequency of meetings depends on the competence of the project team, the complexity of the project, the penalties for missing due dates, and the numbers of people who must be kept informed

Project Team Organization Complete

Determine objectives

Determine the service objectives the network is required to meet. Voice network service is measured in terms of percent blockage and percent of network availability. Determine the required in-service date. Determine growth expectations and the expected life of the service for which the network is being developed.

Identify constraints

Determine from all project team members any constraints they have with respect to force availability, inability to work certain dates, availability of key personnel, or other factors that may affect project completion. Document all such constraints in the minutes.

Develop task list

Develop a detailed list of tasks that must be performed. Assign each task to a responsible individual. Sequence tasks in the order in which they must be performed. Obtain estimates of the amount of time required for each task.

Develop detailed cutover schedule

Identify milestones in the schedule. Set dates for completion of each milestone. Compute the critical path and determine whether the schedule fits within the objective interval. If it does not, determine how to compress the schedule until it fits. Create Gantt and/or PERT charts showing task sequence and schedules, and distribute to all team members.

Project Schedule Complete

Module 2: Develop Requirements

Obtain traffic volumes

Obtain a complete summary of traffic usage measured in minutes of use. Separate international from domestic traffic. International traffic should be summarized by terminating country. Domestic traffic is not distance sensitive in most plans, but where the company has operations in other cities, the point-to-point volume is needed for evaluating the effectiveness of dedicated circuits. Separate traffic into day, evening, and night/weekend.

Determine architecture of existing network

Determine what current networks are in place, and which of these must be retained. Obtain circuit designations, equipment nomenclature, applications, and other factors that could have an impact on the design.

Determine feature requirements

Carriers can offer a variety of special features for both 800/888 and outgoing service. Review the need for such features such as time-of-day routing, network call transfer, forward on busy or don't answer, etc.

Determine constraints

Determine what constraints affect the network design. Examples are existing equipment that must be reused, peak work load times during which no disruption is permitted, etc.

Determine voice/data sharing objectives

Determine the degree to which the network can be shared between voice and data. T-1 lines to the IXC can often be split between voice and data to reduce access circuit costs.

Determine dedicated circuit locations

Determine any locations on the network that can justify dedicated circuits. In making this analysis, review the opportunity for sharing between voice and data. Consider the feasibility of tail-end hop-off (switching calls through the terminating PBX to other destinations such as the local calling area).

Determine commitment levels

The best prices for switched services are obtained by agreeing to term and volume commitments. However, the markets are so volatile that term agreements may put the customer at a disadvantage. Determine the length of term and volumes the company will accept.

Review switching system adequacy

Determine whether the PBXs are capable of carrying the added traffic volume. If not, consider augmenting them. Determine whether trunking must be augmented to carry the added load.

Determine security requirements

Determine provisions that must be programmed into the network to prevent unauthorized access.

Determine node locations

Many voice networks use tandem switching so that hub switches carry traffic destined for other hub locations. Identify the hubbing locations, which are usually chosen for their effect on circuits costs and because they represent major concentrations of usage.

Obtain telephone number list

Obtain a list of the trunk numbers of all services the company has. Include business lines used for fax and modems. Obtain addresses of all services. For PBXs, obtain a complete list of trunk numbers; i.e. any number to which long distance service could be billed. The purpose of this list is to obtain the best quotation and to ensure that the winning carrier has a complete list of numbers to be billed under the agreement.

Obtain credit card details

Determine the number of long distance credit cards assigned within the company. Summarize the amounts and types of calls placed. Determine locations where 800/888 numbers should be installed to reduce calls from the field to company locations.

Determine facility types

Determine what types of facilities will be used for the service. Most voice networks use T-1 for access circuits to the IXC and for connections between other switches.

Determine network management requirements

Determine how the network will be managed. Determine how trouble will be isolated, where alarms will terminate, and what vendors will be responsible for network problems. Determine help desk requirements.

Develop disaster plans

Determine what steps are needed to prevent loss of service during unusual conditions. Determine the need for special 800/888 routing, need for self-healing access circuits, services to alternate carriers' points-of-presence, and other such features that carriers offer to prevent loss of service.

Develop configuration information

Determine information required to configure circuits and equipment. Determine signaling type and whether D-4 or ESF framing will be used on T-1 lines.

Requirements and Specifications Complete

Module 3: Select Equipment

Determine equipment requirements

Determine types and quantities of equipment such as CSUs, additional PBXs or tandem switches, T-1 or T-3 multiplexers, etc. needed to implement the plan.

Estimate circuit costs

Determine circuit costs from carriers. Obtain an estimate of equipment costs from representative vendors.

Evaluate cost/performance characteristics of alternative designs

Review different design alternatives to determine which produces the best combination of price and performance. Detail the differences between alternatives and obtain approval to proceed.

Write equipment request for proposals

Prepare a request for bids or proposals to provide network equipment. Establish a date for required responses. If appropriate, include a requirement for vendor installation and setup. Issue the RFP to selected vendors.

Equipment RFP Responses Received

Review responses

Review vendor responses to the RFP. Reject nonconforming proposals. Rank the remaining proposals in order based on price, references, implementation capability and other such criteria. Check vendors' references. Invite vendors to make product presentations if appropriate.

Select equipment vendor

Select the winning equipment proposal. Negotiate contract terms and conditions with the vendor.

Equipment Selected

Order equipment

Order network equipment from the selected vendors. Establish due dates to conform to the required service date and the date of circuit availability.

Prepare switched service RFP

Write a request for proposals for switched services, including both outgoing and 800/888 for all locations. Provide traffic volumes and commitment levels to the carriers. Set a due date for responses. Include dedicated services, and link to data network requirements if appropriate.

Long Distance Service RFP Responses Received

Review long distance proposals

Open the proposals. Discard the proposals that fail to meet the specifications. Compare responses based on the cost over the life of the contract. Evaluate all cost issues such as signing bonuses, costs of access circuits, penalties, installation charges that are charged or waived, and other such tangible costs. Consider the value of intangibles such as the depth of the provider's support, vulnerability of the network, special reports and billing detail provided, etc. Rank the proposals by value. Pay close attention to the carriers' network architectures and assess them for reliability.

Select service provider

Select the carrier that will provide the service. Review terms and conditions if the proposed contract. Negotiate terms if appropriate.

Sign RESPORG form

The customer signs a form permitting the IXC to act as the responsible

	organization for transferring 800/888 numbers from one carrier to another. This form is required only if IXC is being changed.
Module 5: Plan Cutover	
Plan the project	Meet with the carrier and, if appropriate, the LEC, PBX vendor, and others who have a part in the project. Develop a detailed plan for the remainder of the project. Develop a detailed plan for the remainder of the project including due dates, responsibilities, and reporting requirements.
Determine specific equipment configuration	Determine the quantities of equipment such as CSUs, multiplexers, PBX cards etc. that will be required to implement the system.
Determine specific network configuration	Develop a final configuration concerning major network node locations, circuit quantities, required due dates, etc.
Determine signaling plan	Based on the circuit and equipment selected, determine the signaling arrangements that will be used.
Develop numbering plan	Determine what numbering ranges will be used in each location. Coordinate the numbering plan with DID ranges and extension numbering in each location.
Develop testing plan	Develop a plan for testing all network connections including tests of 800/888

services and abbreviated dialing plans.

Develop synchronization plan

All devices on a digital network must be synchronized. Devices such as PBXs and routers take their timing from the source closest to the national reference frequency. Other devices on the network slave from the master. Be careful to avoid synchronization loops. Prepare a synchronization plan. Consult an expert if in doubt about plan validity.

Develop final security plan

Based on circuits and equipment selected, plan how security will be administered.

Cutover Planning Complete

Module 6: Order and Install Equipment and Circuits

Order circuits

Order circuits from the selected carrier. Set due dates and obtain carrier concurrence that dates can be met.

Order equipment

Order equipment for PBX expansion and network connections from the vendor. Establish due dates corresponding to the required cutover dates.

Circuits and Equipment Ordered

Install equipment

Vendor installs network equipment in all locations. Install equipment in the sequence in which circuits will be turned up. See Chapter 2 for additional details on PBX.

Install access circuits

Carriers install access circuits in all locations. Carriers test circuits end-to-end for point-to-point circuits or from customer locations to the network node for access circuits. CSUs are installed. Carriers loop back the CSUs and test for performance.

Connect access circuits

Access circuits are connected to the PBXs and tested for connectivity to the carrier.

Program ARS

Program the automatic route selection in all PBXs on the network to direct calls to any new trunk groups established as a result of this project. Program the new PIC codes into the ARS for dialing intra-LATA calls.

Test network

Network connections are tested through the equipment end-to-end. Test pairs of termination points.

Module 7: Cutover System

Transfer outgoing services

Where existing outgoing services will be retained, transfer network connections as necessary.

Transfer 800/888 services

Where a change of carriers is involved, the new IXC transfers 800/888 numbers from the old IXC. See Chapters 2 and 3 for additional details.

Cutover Complete

Module 8: Accept System

Evaluate service

Conduct post-cutover service evaluation. Determine whether the system is meeting its objectives. Review statistics from the network to determine whether its usage is meeting expectations. If possible, interview users to determine their satisfaction with the system.

Review carrier bills

Review bills from all carriers involved in the project to ensure that rates are charged according to the contract, that all signing bonuses and usage credits have been applied, that installation charges are properly billed, and that the ARS in all PBXs is routing calls correctly.

Accept network

The customer accepts the network from vendors, makes final payment on equipment, and places the network in service.

Project Complete

Chapter 11
Internetworking Equipment

"We never know how high we are
Till we are called to rise
And then if we are true to plan
Our statures touch the skies"

Emily Dickinson

Internets are the bridge between local and wide area networks. The principal components, routers, switches, and bridges, all have similar planning and implementation requirements. Unlike the other chapters in this book, this chapter is not designed to stand on its own complete with planning, equipment acquisition, installation, and acceptance. The three modules in this project, which are routers, switches, and hubs, are intended to be inserted into WAN or LAN projects and used as part of the master project.

Hubs are not internetworking devices but they are part of nearly all LAN projects. In many LANs they are either being augmented by or replaced by switches, so we have included them in this chapter instead of Chapter 9.

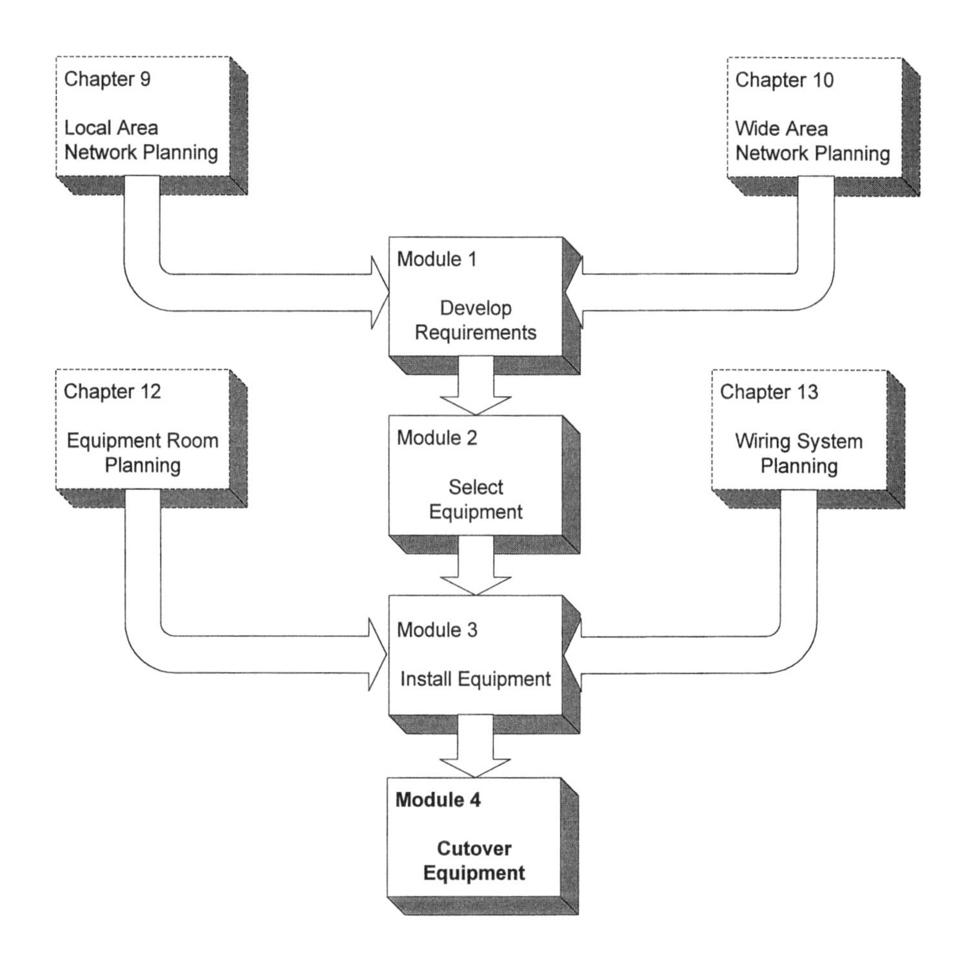

Figure 11-1
Internetworking Equipment
Installation
Project Chart

Hub Installtaion

Module 1: Develop Requirements

Develop hub requirements	Determine the number of hub ports the installation must support. Determine whether to use stackable or modular hubs. Determine the LAN protocols (token ring, Ethernet, fast Ethernet, Appletalk, etc.) that the hub must support.
Determine wiring requirements	Determine whether existing wiring is adequate or whether the premises must be rewired. Determine the number of wire runs to install. Consider quantities of patch cords needed. See Chapter 13 for additional information.
Determine hub topology	Determine where to locate hubs (normally based on where wire is terminated). Consider the need for fiber optics to interconnect hubs.
Determine physical mounting methods	Determine whether hubs will be relay rack-mounted or whether they will be mounted on shelves or fastened directly to the backboard.
Determine network management requirements	Determine level of network management support required. Determine whether hub will be managed or unmanaged, and if managed whether RMON is required. Determine management workstation requirements.
Determine support requirements	Determine value-added features the vendor is expected to bring to the project. Support may include installation, configuration, and ongoing diagnostic and repair support.

Requirements and Specifications Complete

Module 2: Select Equipment

Issue RFP — Issue request for quotations or proposals to select the hub that best meets the requirements.

Evaluate responses — Evaluate the responses to the RFP. Select the winning proposal based on the highest value (not necessarily the lowest price).

Equipment Selected

Develop configuration parameters — For each protocol (IP, IPX, etc.) to be configured on each hub or module, develop the necessary parameters, including addressing requirements.

Order hub — Place an order with the vendor for the selected hubs.

Equipment Ordered

Module 3: Install Equipment

Prepare equipment space — Install all necessary patch panels, conduit, grounds, relay rack, and power outlets to support the hub. (See Chapter 12 for additional details).

Unpack and inventory hardware — The hub is unpacked, inspected for damage, and checked to ensure that all manuals, cables, and other equipment is included.

Mount hub	The hub is bolted into a relay rack or mounted on a shelf as supplied by the customer.
Power on hub	Connect the hub to power and isolated ground. Run all power-on tests prescribed in the manual. Burn in the hub for at least 24 hours.
Connect hub to the backbone and to workstations	If the hub is connected to an internal backbone, install the necessary interface card and connect to the backbone. Install patch cords to connect hub to the workstation wiring.
Configure parameters for all protocols being used	Configure addressing information, subnet mask, SNMP configuration, and other variables involved in the hub configuration.
Test connectivity through the hub	Use the TCP/IP Ping command to test connectivity to other devices on the network.

Module 4: Cutover System

Place hub in service	Connect other LAN devices (servers, routers, etc.) to the hub. Interconnect to other hubs and run LAN acceptance tests. (See Chapter 9 for additional details).
Turn hub over to customer	Provide all manuals, training, record of serial numbers, and other information required in the agreement to the customer.

Cutover Complete

Router Installation

Module 1: Develop Requirements

Develop router requirements — Determine protocols that the router must support. Determine WAN requirements (number of circuits, type of interface, etc.). Determine packet throughput requirements.

Determine network management requirements — Determine level of network management support required. Determine whether router will be managed or unmanaged, and if managed whether RMON is required.

Determine support requirements — Determine value-added features the vendor is expected to bring to the project. Support may include installation, configuration, and ongoing diagnostic and repair support.

Requirements and Specifications Complete

Module 2: Select Equipment

Develop configuration parameters — For each protocol (IP, IPX, etc.) to be configured on each router port, develop the necessary parameters, including addressing requirements.

Issue RFP — Issue request for quotations or proposals to select the router that best meets the requirements.

Evaluate responses — Evaluate the responses to the RFP. Select the winning proposal based on the highest value (not necessarily the lowest price).

Equipment Selected

Order router — Place an order with the vendor for the appropriate router and selected equipment such as CSU/DSU.

Order WAN circuits — Place the necessary orders with the LEC or IEC for wide area network circuits.

Equipment and Circuits Ordered

Module 3: Install Equipment

Prepare equipment space — Install all necessary patch panels, conduit, grounds, relay rack, and power outlets to support the router.

Install WAN circuits — Circuits to connect to the wide area network are installed and tested.

Unpack and inventory hardware — The router is unpacked, inspected for damage, and checked to ensure that all manuals, cables, and other equipment is included.

Mount router — The router is bolted into a relay rack or mounted on a shelf as supplied by the customer. Connect the router to power and isolated ground.

Power on router — Power up the router and run all power-on tests prescribed in the manual. Burn in the router for at least 24 hours.

Connect router to the backbone — If the router is connected to an internal backbone, install the necessary interface card and connect to the backbone. If the router is connected to a WAN, connect the router to the CSU/DSU

Configure parameters for all protocols being used	Configure addressing information, subnet mask, SNMP configuration, and other variables involved in the router configuration. Set up firewall if required.
Test connectivity through the router	Use the TCP/IP Ping command to test connectivity to other devices on the network.

Module 4: Cutover System

Place router in service	Connect other network devices (hub, switches, etc.) to the router. Run network acceptance tests. (See Chapter 9 and 10 for additional details).
Turn router over to customer	Provide all manuals, training, record of serial numbers, and other information required in the agreement to the customer.

Cutover Complete

Switch Installation

Module 1: Develop Requirements

Develop switch requirements	Determine number of ports the switch requires. Determine LAN protocols (10-Base-T, 100-Base-T, token ring, etc.) that the switch must support.

Determine network management requirements — Determine level of network management support required. Determine whether switch will be managed or unmanaged, and if managed whether RMON is required.

Determine support requirements — Determine value-added features the vendor is expected to bring to the project. Support may include installation, configuration, and ongoing diagnostic and repair support.

Requirements and Specifications Complete

Module 2: Select Equipment

Issue RFP — Issue request for quotations or proposals to select the switch that best meets the requirements.

Evaluate responses — Evaluate the responses to the RFP. Select the winning proposal based on the highest value (not necessarily the lowest price).

Equipment Selected

Order switch — Place an order with the vendor for the appropriate switch.

Equipment Ordered

Module 3: Install Equipment

Prepare equipment space	Install all necessary distribution boxes, conduit, grounds, relay rack, and power outlets to support the switch.
Unpack and inventory all hardware	The switch is unpacked, inspected for damage, and checked to ensure that all manuals, cables, and other equipment is included.
Mount switch	The switch is bolted into a relay rack or mounted on a shelf as supplied by the customer.
Power on switch	Connect the switch to power and isolated ground and run all power-on tests prescribed in the manual. Burn in the switch for at least 24 hours.
Connect switch to the backbone and workstations	Connect the switch to the internal backbone, if any. Connect the switch to workstation cabling.
Configure parameters for all protocols being used	Configure addressing information, subnet mask, SNMP configuration, and other variables involved in the switch configuration.
Test connectivity through the switch	Use the TCP/IP Ping command to test connectivity to other devices on the network.

Module 4: Cutover System

Place switch in service	Connect other LAN devices (hubs, servers, routers, etc.) to the switch. Interconnect to other hubs and run LAN acceptance tests. (See Chapter 9 for additional details).

Turn switch over to customer

Provide all manuals, training, record of serial numbers, and other information required in the agreement to the customer's representative.

Cutover Complete

Chapter 12
Equipment Rooms

"All projects should be like our guided missiles

—self destroying"

Robert D. Gilbreath *Winning at Project Management*

Equipment rooms are, strictly speaking, not part of telecommunications, but they are part of virtually every major telecommunications project. In fact, if equipment rooms are not planned and coordinated as part of telecommunication projects, they can easily cause missed due dates or additional expense. We have, therefore, set aside this chapter for equipment room planning. As with some of the other chapters, this one can be cut and pasted into another plan such as PBX, ACD, LAN, etc.

Most of the major activities associated with equipment planning and construction are involved with the building trades. In most companies these are handled by the Facilities Department. Therefore we have limited much of the detail that is not directly involved in telecommunications. Parts of the equipment room plan that are directly related to telecommunications are covered in this chapter. These include such issues as electrical outlets, HVAC, lighting, floor treatment, and security. Although the Facilities Department is responsible for providing the infrastructure, the Telecommunications Department establishes the specifications, and these are the issues covered in this chapter.

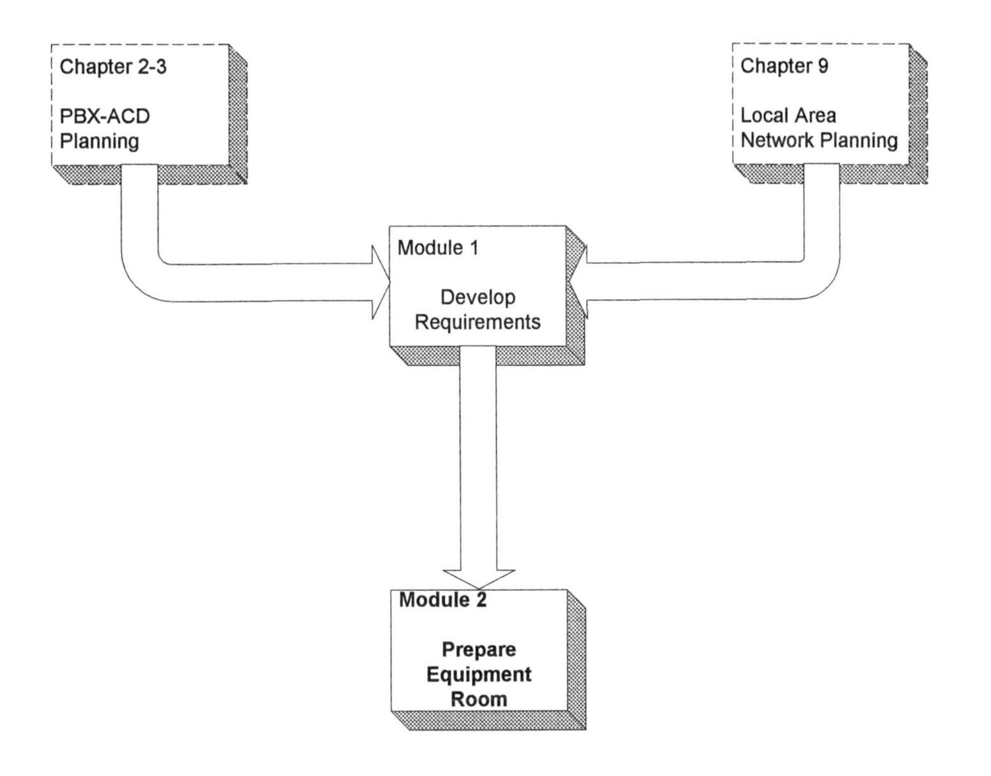

Figure 12-1
Equipment Room Preparation
Project Chart

Module 1: Develop Requirements

Determine equipment room dimensions

The specifications begin with a determination of the dimensions required for the equipment room. Consider all equipment that will be installed in the room including PBX, distributing frames, miscellaneous relay racks, computers, file servers, hubs, routers, network management systems, maintenance terminals, and printers. If appropriate, provide space for a technician's desk. If a wall-mounted MDF will be used, provide space for the backboard. Floor space must be provided for technician access ideally, at least three feet on front and back of equipment. Provide sufficient backboard space for a wall-mounted distributing frame, or for a relay rack-mounted frame. Determine space requirements for other devices such as multiplex, fiber optic terminating equipment, network channel terminating equipment such as repeaters, digital announcers, hubs, routers, and patch panels. Consider floor treatment. Raised floors are ideal, carpets are undesirable.

Determine security requirements

Provide access control though locked doors, card reader systems, etc. Provide smoke and fire alarms. Consider fire suppression systems.

Determine floor loading requirements

Determine the amount of floor loading required for the equipment room. Provide sufficient illumination.

Select equipment room location

Where a choice of location is possible, choose a space close to the wiring center of the building. Choose a location from which the maximum number of stations can be wired within the limits of 90 meters of wire length. Select a space that has adequate security, lighting, heating, ventilation, air conditioning, floor loading, and floor space to contain the planned equipment. Basement spaces should be avoided if possible because of the chance of flooding. Upper floors in wood-frame buildings are undesirable from a fire protection standpoint. Avoid rooms with overhead water pipes because of the potential for equipment-damaging leakage.

Determine electrical wiring requirements

Plan for dedicated power circuits in sufficient quantity to carry the total equipment load, plus leaving a margin for growth. Determine from the vendors how many circuits, of what voltage and amperage, and what type of connectors are required. Check to be sure that the receptacles are located close enough to the equipment. Consider wiring requirements for telecommunications closets. Be certain that the electrical service has enough capacity available. Consult a licensed electrical engineer in case of doubt.

Determine telecommunication s closet requirements

Determine location of telecommunications closets, which may be required to keep wire lengths to 90 meters or less.

Determine grounding requirements

Electronic equipment requires a ground connection tied directly to the building ground. It must be of sufficient size to handle all equipment in the building. It is routed to all satellite equipment rooms and telecommunications closets. Refer to ANSI/EIA/TIA-607, *Commercial Building Grounding and Bonding Requirements for Telecommunications* for details on grounding standards.

Determine heating requirements

All electronic apparatus has operating temperature and humidity limits. Auxiliary heating equipment may be required to keep temperatures up to required minimums during cold weather. All electronic equipment contributes heat to the room, so this should be taken into consideration when calculating heating requirements.

Determine air conditioning and ventilation requirements

Most electronic equipment functions in an environment that is suitable for office workers. Some apparatus, however, has temperature maximums that cannot be exceeded for lengthy periods without damage. Sufficient ventilation must be provided to dissipate heat given off by the equipment. Air conditioning will be required for parts of the country where temperatures rise above the manufacturer's recommended maximum. The vendors can supply heat dissipation figures so the Facilities Department can calculate air conditioning requirements.

Develop pathway requirements

Determine requirements for raceways, conduit, or other pathways from the carriers' point of entrance and between the equipment room and telecommunications closets.

Requirements and Specifications Complete

Module 2: Prepare Equipment Room

Prepare drawings

Prepare floor plans for the equipment room and telecommunications closets. Prepare electrical, illumination, and HVAC drawings.

Plans and Drawings Complete

Prepare equipment room

Install framing, sheet rock, floor and ceiling treatment, doors, and other elements of the equipment room in accordance with the design.

Install electrical wiring

Install dedicated power circuits in the equipment room and telecommunications closets in accordance with the wiring plans.

Install grounding backbone

Install grounding system in accordance with the grounding plan and electrical drawings.

Install heating system

Install heating system in accordance with the heating plan and drawings.

Install air conditioning and ventilation system

Install air conditioning and ventilation system in accordance with the air conditioning plan.

Install security systems

Install access control and alarm system. Install smoke detectors and fire suppression system. Install and connect alarms.

Project Complete

Chapter 13
Wiring Systems

"A little work, a little play,
To keep us going—and so, good day!"

George Louis Palmella

Station wiring projects are common to both data and voice networks, and should be taken into account with every major telecommunications project. Until the last few years, station wiring plans have been developed by the LAN or PBX vendors with no attention to standards. With the completion of the EIA/TIA wiring standards, both manufacturers and users can have uniform plans for station wiring, wiring pathways, equipment rooms and wiring closets, labeling, and bonding and grounding. Although compliance with standards is voluntary, conforming to them ensures compatibility.

This project is divided into separate modules for designing the wiring system, selecting a wiring contractor, installing and accepting the wire, and accepting the project. This project is designed to interface with other projects for installing PBX, ACD, local area networks, and backbone cabling systems. Those other projects have wiring tasks that coordinate with this project. This project can be used by itself coordinated with another project, or cut and pasted into another project.

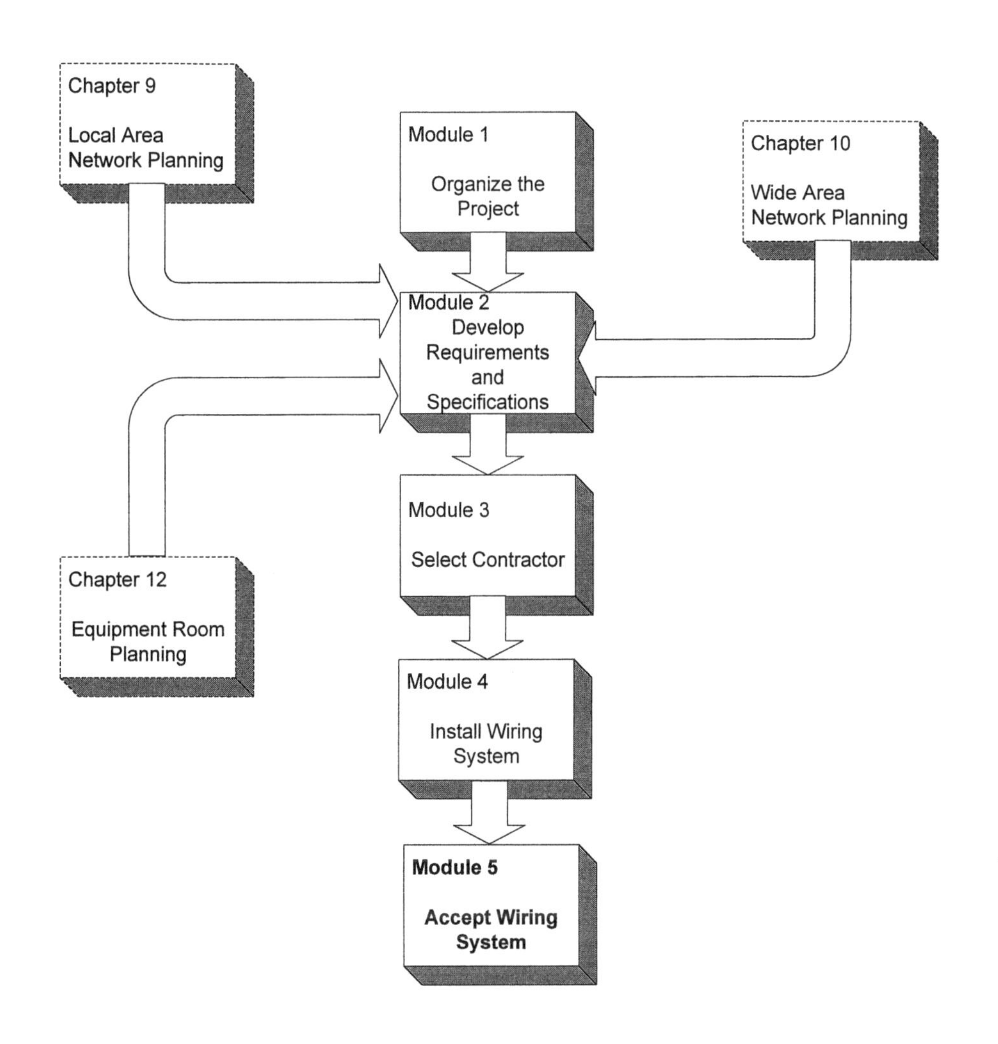

Figure 13-1
Wiring System Installation
Project Chart

Module 1: Organize the Project

Set project objectives

Determine key objectives for the project. These must include completion date plus dates for any intermediate tasks that may affect target completion dates. Determine budgets for the project. Determine what other projects this one is coordinating with, and their key dates.

Organize project team

Identify all who are needed in the project. Representatives should be identified from internal departments that must be represented. Typically, these include the telecommunications, information systems, and facilities departments. Be certain that roles and responsibilities of all team members are clearly understood and accepted.

Hold kickoff meeting

In the initial project meeting all team members should understand their own and others' responsibilities. The kickoff meeting has the following objectives: assign and accept key responsibilities, communicate objectives and constraints of the project, establish schedules and content of reports and project meetings. Someone should be assigned responsibility for preparing minutes and distributing them within a day or two of the meeting.

Set completion dates

Determine key dates in the project, including the final completion dates as well as any interim dates that must be met, such as the date of furniture arrival. If required by a coordinating project, it may be necessary to phase the wiring project by areas of the building.

Schedule project team meetings

If necessary, develop a schedule and place for future team meetings. The frequency of meetings depends on the competence of the project team, the complexity of the project, the penalties for missing due dates, and the numbers of people who must be kept informed. The wiring project may be part of a larger project, and participation included in that team's meetings.

Project Team Organization Complete

Identify constraints

Determine from all project team members any constraints they have with respect to force availability, inability to work certain dates, availability of key personnel or test equipment or other factors that may affect project completion. Document all such constraints in the minutes

Develop task list

Develop a detailed list of tasks that must be performed. Assign each task to a responsible individual. Sequence tasks in the order in which they must be performed. Obtain estimates of the amount of time required for each task.

Develop detailed schedule

Identify milestones in the schedule. Set dates for completion of each milestone. Determine the critical path and determine whether the schedule fits within the objective interval. If it does not, determine how to compress the schedule until it fits.

Create Gantt and/or PERT charts showing task sequence and schedules, and distribute to all team members.

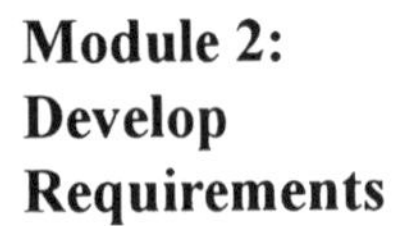

Project Schedule Complete

Module 2: Develop Requirements

Determine wiring requirements

The customer decides on the number and type of drops to be installed to each work station. Determine the EIA/TIA category. Determine whether plenum, or nonplenum wire is required. Determine the need for other types of wire such as shielded or coaxial cable, and for the need for fiber optics to the work area.

Determine wiring methods

Determine how wiring will be routed from the main or intermediate disturbing frame to the work station. Alternatives are poke-through from under the floor, use of floor cells, wiring through suspended ceiling space or attics, and use of surface-mounted raceway. If the poke-through method is used, determine how wiring will be terminated at the work station. Choices are floor monuments or modular furniture channels. For attic or suspended ceiling wiring, determine whether wiring will be fished through walls, use existing conduit, routed through surface-mounted raceway, or installed in power poles.

Determine wiring specifications

Most wiring systems are installed to EIA/TIA 568 specifications, and detailed specifications are unnecessary. Determine whether EIA/TIA specifications are acceptable or whether they must be modified to meet special requirements of the company.

Develop wire termination specifications

Unless otherwise stated, all wire is assumed to be terminated in RJ-45 outlets. Determine whether standard faceplates will be required or whether faceplates should match the number of outlets to be installed. Determine color standards for jacks. Determine whether keyed outlets will be required for some types of service. Determine how outlets will be designated.

Determine outlet numbering and labeling plan

Determine how outlets will be numbered and labeled at the work station, telecommunications closets, and main distributing frame. Labeling may follow TIA/EIA 606 standards.

Requirements and Specifications Complete

Determine main distributing frame location

The MDF will normally be mounted in the equipment room, but in some installations the equipment will be mounted separately and connected to the MDF. with a tie cable. The MDF must have enough space to terminate all voice wire, data wire, tie cables, PBX ports, fiber optic cable to telecommunications closets, riser cable, and LEC interface blocks.

Design main distributing frame

Space should be designated for all cables and termination blocks that will be placed on the MDF. Provide space for future expansion. Provide wire organizers and cable rings. Determine whether the MDF will be wall- or relay rack-mounted. If the latter, determine whether the MDF will be single- or double-sided.

Select intermediate distributing frame locations

Normally, if IDFs can be avoided, administration is simplified. In many buildings, however, IDFs are required to maintain wire runs within the limits allowed by the specifications. Choose IDFs within 90 wire-meters of workstations. Provide sufficient backboard space to terminate telephone station wire, data station wire, riser cables, and any necessary active network equipment. See Chapter 12 on equipment room planning.

Determine riser cable requirements

In buildings where the LEC's point-of-presence is located away from the equipment room, and where telecommunications closets are used, riser cables are required. Campus riser between buildings also must be considered. Riser pairs should stay within the length limitations specified in EIA/TIA 568. Pairs should be furnished in sufficient quantity to meet current needs with future expansion in mind. A good rule of thumb is to provide at least four pairs per work station. Sheath composition and gauges are selected in conformance with EIA/TIA 568 and appropriate fire protection regulations.

Determine voice cable termination methods

Voice cables are terminated on the main and intermediate distributing frames. Punch-down blocks of the 110- or 66-type or proprietary types from any several manufacturers can be used. For consistency, voice cables should be terminated on the same type of block as the PBX ports.

Determine data cable termination methods

In most installations data wiring will be terminated on RJ-45 patch panels. Determine whether patch panels will be mounted on relay racks or surface-mounted on the MDF and IDFs. Wall-mounted relay racks can be mounted on backboards if sufficient space has been provided. Consider space requirements for hubs. Design relay racks to keep patch cord lengths reasonable. Provide patch cord organizers where necessary.

Determine pathway requirements

Check that sufficient pathway capacity is available to meet cabling requirements. Determine whether new pathways such as conduits, raceways, cable trays, etc. must be added with this project. Determine requirements for lashing cables to the trays and for separation from existing cables. Determine whether sufficient conduit exists for installing risers and station cables, or whether conduit must be reinforced to provide added capacity.

Prepare floor plan

A floor plan showing the location of each wire termination is required. If wire is to be run in modular furniture channels, show the furniture layout. Otherwise, show the location of floor monuments, wall-mounted outlets, and/or power poles.

Floor Plans Complete

Determine due date requirements

Determine the dates on which wiring installation can begin, and the date by which it must be finished. Start dates are usually determined by space availability and ending dates by the move-in date.

Determine warranty requirements

Most wiring contractors warrant their work for a specified period of time. Most manufacturers warrant the materials for a longer period, but a material-only warranty is of limited value since most of the cost of the installation is in labor. Consider a manufacturer's extended performance warranty, in which the manufacturer guarantees performance up to a particular level such as 100 mb/s or 155 mb/s. This type of warranty protects the customer against the contractor's going out of business before the warranty expires.

Determine testing requirements

Determine what types of tests must be performed on the wire. Typically, category 3 cable is tested for shorts, opens, grounds, and turnovers. Category 5 cable is tested with a level 2 test set to prove that the installation falls within EIA/TIA 568 specifications. The contractor should be required to perform all tests and to furnish a copy of the results in printed form or on floppy disk.

Develop evaluation criteria

Determine how the winning proposal will be selected. Price is always a factor, but other factors should be considered such as references from users with similar configurations and ability to complete the project on time.

Module 3: Select Wiring contractor

Prepare bid or RFP documents

Prepare documents describing requirements and specifications so wiring contractors can prepare bids or proposals. Set a due date for responses. Be specific about requirements and specifications. Include due dates, testing and construction requirements, documentation requirements, and requirements for plenum wire. Include a detailed floor plan showing the quantity of drops to be terminated at each location.

RFP Responses Received

Review bids and proposals

Review wiring contractors' bids and proposals. Select the winning contractor. Consider factors other than price such as references, contractor stability, ability to complete on schedule, and other such variables.

Prepare agreement with selected wiring contractor

Develop a written agreement with the selected contractor that interprets the terms of the customer's request and the contractor's response. Incorporate important points such as due dates, product to be used, warranty, and other such items that should be covered by written agreement. On contract signing, the customer will be expected to pay an initial amount, which may cover the cost of the materials.

Contractor Selected

Module 4: Install Wiring

Obtain permits

Obtain permits from governmental bodies that regulate cable placement.

Order materials

The contractor places the order for wiring materials.

Deliver materials

Materials are delivered to the job site and inventoried.

Materials on Site

Remove old wire

If existing wire must be removed, it is removed before wiring work begins.

Prepare and install backboards

Backboards for the equipment room and telecommunications closets are cut to size, treated with fire-resistant paint, and mounted securely to the wall.

Install relay racks

Any wall-mounted or floor-mounted relay racks required for mounting patch panels and network equipment are installed.

Install cable rings and wire organizers

Cable rings for holding jumpers and cables are installed on the backboards.

Install pathways

Any conduits, raceways, or cable trays required for the project are mounted.

Install riser cable

Riser cable is pulled between distributing frames, fastened securely, and terminated on blocks.

Deliver furniture

Where wiring is being placed in modular furniture, the furniture is delivered in time to meet the installation schedule.

Furniture on Site

Install station wire

Station wire is pulled through the pathway from the main or intermediate distributing frames to the work area. Wire is suspended from the building's structure as specified in EIA/TIA 568.

Terminate wire

Category 3 wire is terminated on punch-down blocks at the distributing frame end. Category 5 wire is terminated in patch panels. Wire is terminated on RJ-45 jacks at the station end unless otherwise stated in the specification.

Test wire

Category 3 wire is tested on a DC basis. Category 5 wire is tested for full compliance with EIA/TIA 568 requirements. DC tests are made on riser cables. Test results are provided to the customer.

Label wiring

Install all labeling on wiring as required by the specifications.

Prepare as-built drawings

The building plans are modified as necessary to show the final plan as left by the contractor.

Perform quality inspections

The wiring project is inspected to verify compliance with the specifications. All cables are checked for proper dress. A random sample of Category 5 telecommunications outlets are inspected for proper termination and lead twist. Labeling is inspected for compliance with the specifications. If governmental agencies require inspection, have the project inspected for code compliance. Inspector's approval is required for acceptance of the project.

Installation Complete

Module 5: Accept System

Review as-built drawings

Contractor provides as-built drawings as required by the specifications. Customer reviews the drawings for accuracy.

Review test results

Review the documentation the contractor provides on wire and fiber optics. Check that all runs are within specifications and that all pairs have been tested.

Accept project

When all inspections are complete, test documents have been provided, and as-built drawings provided, the project is accepted and final payments become billable.

Project Complete

Chapter 14

Backbone Cabling and Conduit Systems

"A pint of sweat saves a gallon of blood"

General George S. Patton

Backbone Conduit

When exterior backbone cables are being placed, it is always advisable to consider installing them in conduit. Aerial cable is susceptible to storm damage, and once cable is buried, it is in the ground for good. Conduit offers mechanical protection and makes it relatively easy to replace cable if a larger size is needed or if the cable is damaged. Even if you plan to bury cable, much of the cost is in opening the trench. While the trench is open, consider leasing extra conduit in case it is needed in the future. Whenever roads, driveways, and sidewalks are installed in a campus environment, consider installing conduits or sleeves under the pavement in case cable must be placed later.

In new buildings, a conduit to the public right-of-way is required for common carriers to install their facilities. Consider placing extra conduit in case you want to add service from a different carrier some time in the future. Also, consider installing cable to a second entrance location to provide diversity if it is needed in the future.

Conduit is sometimes required inside buildings to provide mechanical protection or to contain a cable sheath that fire codes prohibit in exposed areas. This section discusses the elements of a conduit project. The plans in this chapter assume that a project team is not needed, and that most of the planning is done by an informally organized group.

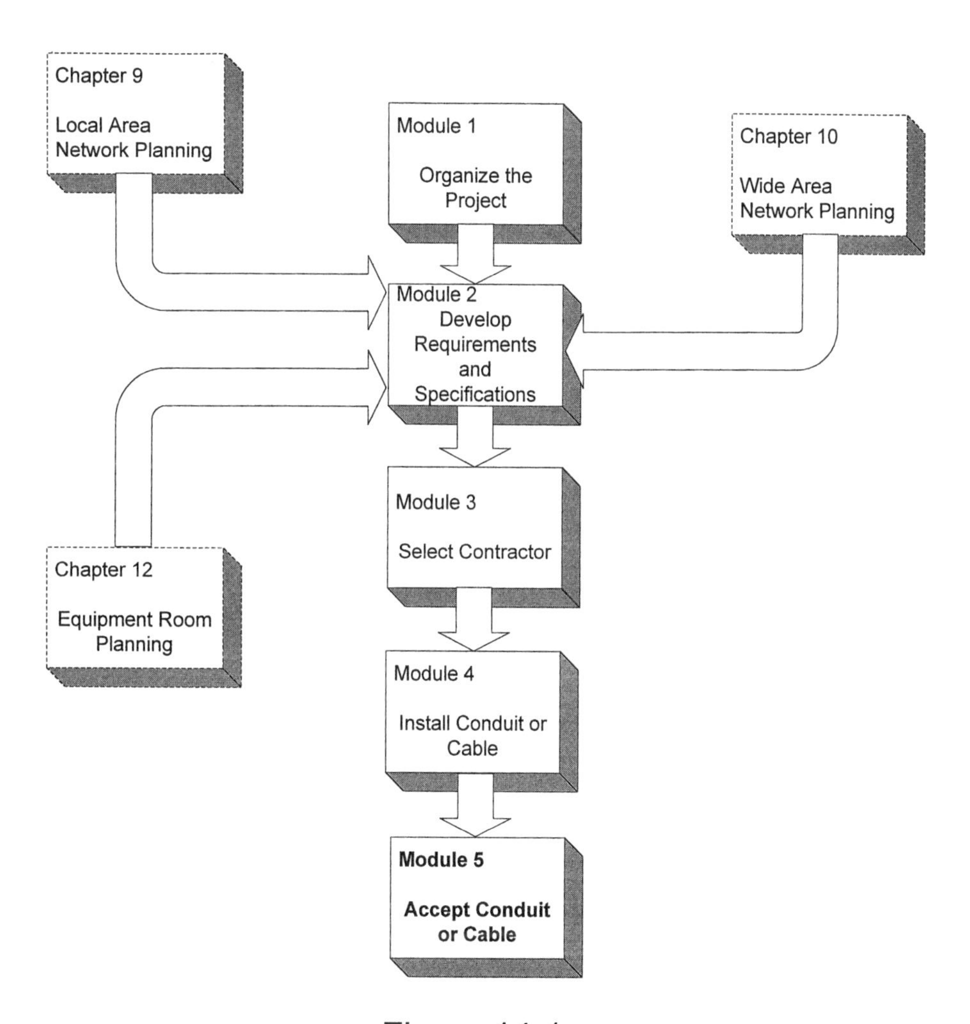

Figure 14-1
Backbone Cable and Conduit Installation
Project Chart

Module 1: Organize the Project

Determine project objectives

Determine budget expectations and the required completion date for the project. Determine capacity requirements and expected growth.

Identify constraints

Identify any constraints such as the owner's policy on pavement cuts, any landscaped areas that cannot be disturbed, times of the day when machinery noise cannot be tolerated, and other such constraints. Determine any special building code requirements imposed by the jurisdiction in which the conduit will be installed.

Project Organization Complete

Module 2: Develop Requirements

Determine conduit route

Determine the route the conduit will follow between termination points. Measure the footage required.

Determine installation method

Outdoor conduit is usually trenched or some contractors can bore under paved areas using guided boring tools. Contractors can also force conduit under paved areas by digging a pit and pushing a pipe across with a hydraulic tool.

Determine the size of cable required

The size of the conduit is determined by the size of the cable it must support. Cable manufacturers can provide charts to show the capacity of different sizes of conduit.

Determine innerduct requirements

Conduits are often subdivided with innerduct, which is a flexible sub-duct that can be pulled in conduit to protect fragile cables such as fiber optics, and to provide for future expansion. Determine the size and number of sub-ducts that will be placed.

Determine conduit type

The major types of conduit used in campus environments are PVC for exterior use and EMT for interior use. Determine which type will be used in the project.

Determine access hole requirements

When conduit runs are too long to permit installers to pull cable in one continuous run, when cable must be spliced, or when branches or 90 degree bends are required, access holes are advisable. These range from large underground vaults or manholes to small hand holes that are accessible from ground level. Determine the location of access holes and the size required. Determine whether lockable covers are required.

Determine size and quantity of conduit

PVC conduit is available in two inch, three inch, and four inch diameters. Metallic conduit is available in sizes as small as 1/2 inch. Determine the diameter required based on the cable diameter. Determine how many conduits should be placed. Consider placing additional conduit for future growth.

Requirements and Specifications Complete

Prepare construction drawings

Prepare drawings showing location and type of all conduits to be placed. Identify all locations where conduit must cross paved areas. Identify buried utilities. When necessary, have specialists locate these.

Prepare job specifications

Develop a specification detailing the size and type of conduit to be placed, the depth to which it is to be buried, the method of fixing interior conduits to the building structure, the required completion date, etc.

Job Drawings and Specifications Complete

Module 3: Select Contractor

Issue bid requests

Issue the specification and drawings to qualified contractors to prepare bids. Consider requesting separate prices for installation and materials as an aid to comparing responses. Establish a due date for responses.

Evaluate bids

Open the bids from respondents and compare them. Determine capabilities of low bidders based on references and experience with similar projects. The vendor is selected and a contract is written.

Contractor Selected

Module 4: Install Conduit

Obtain permits Obtain permits from governmental bodies that regulate conduit placement.

Order materials The contractor orders materials and delivers them to the site.

Open trenches If the conduit is being placed in the ground, the contractor bores or digs trenches as appropriate.

Place conduit Install conduit in trenches or through bored openings. Interior conduit is routed through specified areas and fastened to the building structure.

Backfill trench The contractor backfills the trench and restores landscaped areas.

Installation Complete

Module 5: Accept Project

Inspect the project The customer inspects the work done. If governmental agencies require inspection, have the project inspected for code compliance. Inspector's approval is required for acceptance of the project. On direct buried cable, either observe the placement job or dig down to the cable in selected locations to verify the depth of burial. The contractor is given a punch list of items needing correction.

Accept project. The contractor completes the punch list. The owner accepts the project and pays final bills.

Copper cable

Copper cable is used extensively in campus environments as riser cable between buildings, within buildings as riser between floors, between the equipment room and the carrier's point-of-presence, and between the equipment room and telecommunications closets. Despite the greater bandwidth and noise immunity of fiber optics, copper cable is still the most practical medium for connecting voice circuits to PBXs and the public network.

Backbone cable techniques are usually different than station wiring, which is covered in Chapter 13. Station wiring is constructed to rigid specifications to enable it to carry high-bandwidth data. Backbone cables are intended for voice. Although backbone cables can and do carry data, the bandwidth is limited and the crosstalk performance is lower.

This section discusses considerations for planning and installing copper cable. Most cables will be installed in conduit using the techniques discussed in that section. If the cable is to be directly buried, refer to the conduit section for considerations when selecting the route, trenching, and backfilling.

Module 1: Organize the Project

Determine project objectives Determine the purpose or which the cable is required. Determine project budgets. Determine growth expectations and the date by which the cable must be installed.

Identify constraints Identify any constraints such as policies against the use of exposed cable, local code requirements with respect to cable and sheath types, lengths and gauge restrictions required by equipment, etc.

Project Organization Complete

Module 2: Develop Requirements

Determine size requirements Copper cable is available in sizes ranging from one or two pairs to 3,600 pairs. For riser cable connecting the MDF to telecommunications closets, four pairs per work station is a good rule of thumb. For riser cable from the point-of-presence to the equipment room, one pair per station is usually sufficient. Individual cases may differ widely. For example, when Centrex is used, more pairs are needed. Determine the cable gauge--24 gauge is usually a good compromise. The number of pairs and the gauge may be limited by the size of available conduit.

Determine cable route Determine the route cable will take and measure the footage. Allow extra footage for splices if these are required.

Determine protection requirements. If the cable is exposed to lightning or power cross, protectors are required. Determine whether the cable is exposed (if in doubt consult an expert). Determine the quantities and type of protectors to use.

Determine cable makeup Determine the type of sheath required. PVC is usually used for exterior cable, but cannot the used for extended distances inside buildings. Check plenum requirements for interior cables. Determine depth of any buried cable.

Requirements and Specifications Complete

Prepare detailed specifications

Prepare a drawing showing cable routes, length, measurements to various landmarks, and any placing and splicing requirements. Show termination points and termination methods. Prepare a written specification detailing due dates, placing instructions, and material specifications.

Job Drawings and Specifications Complete

Module 3: Select Contractor

Prepare bid documents

Prepare a request for bids or proposals. Issue to qualified contractors.

Evaluate bids

Open the bids from respondents and compare them. Determine capabilities of low bidders based on references and experience with similar projects. The vendor is selected and a contract is written.

Contractor Selected

Module 4: Install Cable

Obtain permits Obtain permits from governmental bodies that regulate cable placement.

Order materials The contractor orders materials and delivers them to the job site.

Prepare cable pathways Open any trenches required. Place conduit, cable trays, or other pathways in preparation for cabling.

Install cable The contractor places cable in the pathways. The terminating blocks on distributing frames are mounted on back boards or relay racks in preparation for termination.

Splice cable Any splices required are completed including splices to connect protector blocks to the cabel.

Test cable Test all cable pairs for DC faults (opens, shorts, crosses, and grounds). Record results and furnish to the customer.

Installation Complete

Module 5: Accept Project

Inspect the project The customer inspects the work done. If governmental agencies require inspection, have the project inspected for code compliance. Inspector's approval is required for acceptance of the project. On direct buried cable, either observe the placement job or dig down to the cable in selected locations to verify the depth of burial. The contractor is given a punch list of items needing correction.

Accept project. The contractor completes the punch list. The owner accepts the project and pays final bills.

Fiber Optics

Fiber optics is a high-quality, moderate-cost medium for data communications. Most local. Area networks include at least some fiber optics for linking hubs. Some companies are extending fiber to the desktop in preparation for high speed protocols such as ATM. The techniques for placing fiber are similar to those for copper cable with certain exceptions. For one thing, fiber is more fragile, and can be damaged more easily when it is being pulled through conduits. It is subject to bending radius rules to avoid microbending, which increases loss. It is also more difficult to splice: in fact splicing should be avoided altogether if possible.

Module 1: Organize the Project

Determine project objectives Determine the purpose and the objectives of the project. Determine the protocols the fiber is expected to support. Determine budgets and due dates. Determine growth expectations.

Determine constraints Determine any constraints on placing the fiber such as routes that must be used or avoided, specific manufacturers or contractors the customer prefers to use or avoid, etc.

Project Organization Complete

Module 2: Develop Requirements

Determine architecture

Fiber is usually placed either point-to-point or in a ring configuration. Determine which architecture will be used.

Determine size of fiber required

Fiber cables can be obtained with as few as one pair of fibers to as many as 144 fibers in one sheath. Determine the number of pairs needed considering both initial and growth requirements.

Determine fiber mode requirements

Fiber comes as single mode or multimode. Generally, building and campus applications use multimode fiber and common carrier applications use single mode. The type of cable can be determined by consulting the documentation of the equipment manufacturer. Composite cable with both types of fiber are available on special order.

Determine cable sheath requirements

Cables are available with both interior and exterior cable sheath. Interior cable is available with either plenum or nonplenum sheath. Exterior cable is available with a variety of protective sheaths and armoring. It is also commonly available with gel fill to keep water out.

Determine cable length

Select the route for the fiber and measure the amount of cable required. Allow additional footage for coiling excess fiber in cable trays for future splices and rearrangement.

Determine termination methods

A wide variety of fiber connectors is available. Determine from the equipment manufacturer what type of connector is needed. Determine how connectors will be mounted. A common method of mounting is to use a wall-or frame-mounted cabinet with bulkheads for installing connectors.

Determine conduit requirements

Determine the size and type of conduit (PVC, EMT, etc.) required for the project.

Determine innerduct requirements

Innerduct is highly recommended when fiber is sharing conduit runs with copper cable, where a duct is to be subdivided for future expansion with copper or fiber cable, or for identification and protection of fiber in cable trays or suspended ceiling areas.

Determine patch cord requirements

Determine the quantity, length, and connector type required for patch cords between the fiber and the terminating equipment. Determine who will provide and install the patch cords.

Requirements and Specifications Complete

Prepare detailed specifications

Prepare a specification detailing the requirements for the project including the size, type, and mode of fibers required. Indicate whether splicing is permitted or prohibited. Include drawings showing the route to be used if appropriate. Indicate the availability of conduit or innerduct.

Job Drawings and Specifications Complete

Module 3: Select Contractor

Issue bid requests Issue a request for vendors to propose material and installation costs for placing the fiber. Set a due date for response. Include specifications for testing for loss and impedance irregularities.

Evaluate bids Open the bids from respondents and compare them. Determine capabilities of low bidders based on references and experience with similar projects. The vendor is selected and a contract is written.

Contractor Selected

Module 4: Install Cable

Order fiber The customer orders cable from the vendor, who orders it from the manufacture. Order terminating units, splicing trays, and other hardware needed to install and terminate the fiber.

Deliver fiber The manufacturer ships the fiber and the vendor delivers it to the job site.

Obtain permits Obtain permits from governmental bodies that regulate cable placement.

Open trenches If the fiber is being placed in the ground, the contractor bores or digs trenches as appropriate.

Place conduit — Install conduit in trenches or through bored openings.

Place fiber — The installer places the fiber cable in conduit, cable trays, or other enclosures. Splice fiber if required.

Terminate fiber — Terminate fiber in connectors in line interface units.

Test fiber — Contractor tests the fiber for loss and impedance irregularities. A record of the results is furnished to the customer. Any fibers falling outside the accepted range are reterminated and retested.

Installation Complete

Module 5: Accept Project

Inspect project — The customer reviews the results of the tests and inspects the quality of the work done. If governmental agencies require inspection, have the project inspected for code compliance. Inspector's approval is required for acceptance of the project. A punch list is made of any irregularities that must be corrected.

Accept project. — The contractor completes the punch list. The owner accepts the project and pays final bills.

Appendix A

PBX-ACD Installation Acceptance Checklist

In the checklist below, OK means the item was inspected and found to be installed according to the plan. EX means an exception was found, which should be detailed on the reverse side. NA means the item is not applicable.

1.	**Cable Trays and Racking**	OK	EX	NA
1.1.	Superstructure installed at the correct height	☐	☐	☐
1.2.	Cable rack and troughs secured and properly spaced	☐	☐	☐
1.3.	Racks installed with sufficient overhead clearance	☐	☐	☐
1.4.	Cable rack and troughs straight	☐	☐	☐
1.5.	Cable rack and troughs properly terminated	☐	☐	☐
1.6.	Painting and touch-up applied as needed	☐	☐	☐
1.7.	Labeling complies with EIA/TIA 606	☐	☐	☐
1.8.	As-built drawings provided	☐	☐	☐

2.	**Backboards and Wall-Mounted Distributing Frames**	OK	EX	NA
2.1.	Backboard painted with fire retardant paint	☐	☐	☐
2.2.	Backboard fastened securely to wall	☐	☐	☐
2.3.	Backboard segregated by type of connection	☐	☐	☐
2.4.	Cable rings properly installed	☐	☐	☐
2.5.	Terminal blocks properly installed	☐	☐	☐
2.6.	Labeling complies with EIA/TIA 606	☐	☐	☐
2.7.	Ground bus tight and properly labeled	☐	☐	☐
2.8.	Protectors properly installed	☐	☐	☐
2.9.	Lighting and receptacles operational	☐	☐	☐
2.10.	Cables properly secured on verticals	☐	☐	☐
2.11.	Cable and wire free of damage	☐	☐	☐
2.12.	Wire properly terminated	☐	☐	☐

2.13.	Cable performance tests completed per EIA/TIA 568	☐	☐	☐
2.14.	As-built drawings provided	☐	☐	☐

3.	**Free-Standing Distributing Frames**	OK	EX	NA
3.1.	Vertical members plumb and tightly fastened to the floor	☐	☐	☐
3.2.	Cable rack and trough properly secured to frame	☐	☐	☐
3.3.	Ground bus tight and properly labeled	☐	☐	☐
3.4.	Cable rings properly installed	☐	☐	☐
3.5.	Terminal blocks properly mounted	☐	☐	☐
3.6.	Labeling complies with EIA/TIA 606	☐	☐	☐
3.7.	Protectors properly installed and grounded	☐	☐	☐
3.8.	Frame and aisle lighting and receptacles operational	☐	☐	☐
3.9.	Painting and touch-up applied as needed	☐	☐	☐
3.10.	Cables properly secured on verticals	☐	☐	☐
3.11.	Cable and wire free of damage	☐	☐	☐
3.12.	Wire properly terminated	☐	☐	☐
3.13.	Cable performance tests completed per EIA/TIA 568	☐	☐	☐
3.14.	As-built drawings provided	☐	☐	☐

4.	**Relay Rack Equipment**	OK	EX	NA
4.1.	Relay rack equipment mounted per specifications	☐	☐	☐
4.2.	Terminal blocks properly mounted and labeled	☐	☐	☐
4.3.	Equipment covers properly installed	☐	☐	☐
4.4.	Fuses installed, tight, and of proper size	☐	☐	☐
4.5.	Spare fuses installed in fuse holders	☐	☐	☐
4.6.	Alarms tested and operational	☐	☐	☐
4.7.	Equipment tested for proper operation	☐	☐	☐
4.8.	Cables properly secured with no damage	☐	☐	☐
4.9.	Cable conductors properly terminated	☐	☐	☐
4.10.	Cable continuity tests completed	☐	☐	☐
4.11.	Jumpers properly run and dressed	☐	☐	☐
4.12.	Cable ties properly run and dressed	☐	☐	☐
4.13.	Labeling complies with EIA/TIA 606	☐	☐	☐
4.14.	Painting and touch-up applied as required	☐	☐	☐
4.15.	As-built drawings provided	☐	☐	☐

5.	**Power and Ground**	OK	EX	NA
5.1.	Grounding system complies with EIA/TIA 607	☐	☐	☐
5.2.	All fusing correct size and properly installed	☐	☐	☐
5.3.	Spare fuse holders installed with spares provided	☐	☐	☐
5.4.	Isolated ground measured	☐	☐	☐
5.5.	Caution signs installed	☐	☐	☐
5.6.	Painting and touch-up applied as required	☐	☐	☐

5.7.	Labeling complies with EIA/TIA 606	☐	☐	☐
5.8.	As-built drawings provided	☐	☐	☐

6.	**Battery/UPS Installation**	OK	EX	NA
6.1.	Battery racks properly installed	☐	☐	☐
6.2.	Rectifiers level and mounted securely to wall	☐	☐	☐
6.3.	Bus bars installed in accordance with specifications	☐	☐	☐
6.4.	Bus and power connections clean, tight, and cool to touch	☐	☐	☐
6.5.	Meters calibrated	☐	☐	☐
6.6.	Alarms tested and functioning properly	☐	☐	☐
6.7.	Power supply load-tested	☐	☐	☐
6.8.	Battery cell connections clean and tight	☐	☐	☐
6.9.	Battery tests completed and recorded	☐	☐	☐

7.	**PBX-ACD Equipment**	OK	EX	NA
7.1.	Cabinets level, plumb, and fastened to the floor	☐	☐	☐
7.2.	Cabinet doors unobstructed	☐	☐	☐
7.3.	Earthquake bracing properly installed	☐	☐	☐
7.4.	Cabling formed properly	☐	☐	☐
7.5.	Cable conductors properly terminated	☐	☐	☐
7.6.	Network synchronization verified	☐	☐	☐
7.7.	Alarms tested and operational	☐	☐	☐
7.8.	Ringing and tones supplies tested and operational	☐	☐	☐
7.9.	Painting and touch-up applied as necessary	☐	☐	☐
7.10.	Cabinets properly labeled	☐	☐	☐
7.11.	Premises free of damage resulting from installation	☐	☐	☐
7.12.	Premises clean with trash and excess material removed	☐	☐	☐
7.13.	Initialization record sheets complete and on premises	☐	☐	☐
7.14.	Inventory of all cards and equipment	☐	☐	☐
7.15.	Manuals and as-built drawings on premises	☐	☐	☐
7.16.	Transmission tested on all analog trunks	☐	☐	☐
7.17.	Manufacturer's installation tests complete	☐	☐	☐

8.	**System Diagnostics Completed**	OK	EX	NA
8.1.	Tone and ringing supplies	☐	☐	☐
8.2.	Trunks and signaling	☐	☐	☐
8.3.	T-1 systems error-free performance	☐	☐	☐
8.4.	Voice mail	☐	☐	☐
8.5.	Interactive voice response	☐	☐	☐
8.6.	Call management system and associated I/O equipment	☐	☐	☐
8.7.	Common equipment (DTMF registers, etc.)	☐	☐	☐
8.8.	MAT terminal and associated I/O equipment	☐	☐	☐
8.9.	Conference circuits	☐	☐	☐

8.10.	Call detail recorder	☐	☐	☐

9.	**Security Provisions Applied**	OK	EX	NA
9.1.	Class-of-service restrictions	☐	☐	☐
9.2.	Code restriction	☐	☐	☐
9.3.	Maintenance port security device	☐	☐	☐
9.4.	Toll restriction	☐	☐	☐
9.5.	Blocked country/area codes	☐	☐	☐
9.6.	Unassisted transfer through voice mail blocked	☐	☐	☐
9.7.	Trunk class-of-service set according to plan	☐	☐	☐
9.8.	DISA disabled	☐	☐	☐
9.9.	Default password in maintenance terminal reset	☐	☐	☐
9.10.	Default password expiration in voice mail	☐	☐	☐
9.11.	Voice mail minimum password length	☐	☐	☐
9.12.	Transfer through auto attendant menu blocked	☐	☐	☐
9.13.	Vector or routing scripts verified	☐	☐	☐
9.14.	Switchroom access controlled according to plan	☐	☐	☐
9.15.	Trunk access codes restricted	☐	☐	☐

10.	**Verification of Features**	OK	EX	NA
10.1.	Automated attendant routing	☐	☐	☐
10.2.	Automatic call distribution	☐	☐	☐
10.3.	Automatic route selection	☐	☐	☐
10.4.	Call-by-call service selection	☐	☐	☐
10.5.	Call coverage paths	☐	☐	☐
10.6.	Call distribution recorder	☐	☐	☐
10.7.	Call forward - busy	☐	☐	☐
10.8.	Call forward - no answer	☐	☐	☐
10.9.	Computer-telephony integration	☐	☐	☐
10.10.	Data transmission	☐	☐	☐
10.11.	Direct inward dialing	☐	☐	☐
10.12.	Direct outward dialing	☐	☐	☐
10.13.	Emergency transfer	☐	☐	☐
10.14.	Flexible numbering	☐	☐	☐
10.15.	Line hunting	☐	☐	☐
10.16.	Music-on-hold	☐	☐	☐
10.17.	Networking software	☐	☐	☐
10.18.	Paging access	☐	☐	☐
10.19.	Power fail transfer	☐	☐	☐
10.20.	Private network access	☐	☐	☐
10.21.	Remote administration	☐	☐	☐
10.22.	Station-to-station calling	☐	☐	☐
10.23.	System recovery from power loss	☐	☐	☐

10.24.	Tandem switching	☐	☐	☐
10.25.	Tie trunks	☐	☐	☐
10.26.	Traffic measurement	☐	☐	☐
10.27.	Trunk answer from any station	☐	☐	☐

11.	**Verification of Trunks**	OK	EX	NA
11.1.	Incoming call on all C.O. trunks, tie trunks, or PRI groups	☐	☐	☐
11.2.	Outgoing call on all C.O. trunks, tie trunks, or PRI groups	☐	☐	☐
11.3.	Verification of routing of each 800 number	☐	☐	☐
11.4.	Transmission measurements on each analog trunk	☐	☐	☐
11.5.	Busy indication on loss of T-1	☐	☐	☐
11.6.	Incoming call on each DID trunk	☐	☐	☐
11.7.	Paging access tested for proper operation	☐	☐	☐

12.	**Attendant Console and Station Verification**	OK	EX	NA
12.1.	Console operations	☐	☐	☐
12.2.	Centralized attendant service	☐	☐	☐
12.3.	Multiple console overflow	☐	☐	☐
12.4.	Night service arrangements	☐	☐	☐
12.5.	Station set inventory	☐	☐	☐
12.6.	Class of service and assigned features	☐	☐	☐
12.7.	Call originating and completion from each station	☐	☐	☐
12.8.	Intercepted numbers	☐	☐	☐

13.	**ACD Verification**			
13.1.	ACD Supervisory terminals	☐	☐	☐
13.2.	ACD overflow	☐	☐	☐
13.3.	Conditional routing scripts and vectors	☐	☐	☐
13.4.	Skill-based routing	☐	☐	☐
13.5.	Queue announcements	☐	☐	☐
13.6.	Readerboard display	☐	☐	☐
13.7.	Call management system reports	☐	☐	☐
13.8.	Service observing equipment	☐	☐	☐
13.9.	Predictive dialer	☐	☐	☐

14.	**Documentation Provided**	OK	EX	NA
14.1.	Maintenance manuals	☐	☐	☐
14.2.	Software manuals	☐	☐	☐
14.3.	Schematic drawings	☐	☐	☐
14.4.	Backup program tape	☐	☐	☐
14.5.	Battery records	☐	☐	☐
14.6.	Trouble reporting log	☐	☐	☐

14.7.	Cable assignment records	☐	☐	☐
14.8.	Trunk assignment records	☐	☐	☐
14.9.	Line assignment records	☐	☐	☐

Appendix B

Security Checklist

Toll fraud as gained a lot of attention in the past few years as groups of toll thieves have learned how to dial into a PBX and out through an outgoing trunk. Many companies have been hit for tens of thousands of dollars of toll fraud, mostly to overseas locations. In response, PBX manufacturers have taken counter measures that can practically eliminate toll fraud, but the vendor still must install the system properly.

Most call thieves operate by one of three major methods:

- Transferring to an outgoing trunk through voice mail or auto attendant
- Using remote access
- Hacking the maintenance port password and removing barriers

Of the three methods, the last is the most insidious. If a toll thief is able to dial into the maintenance port, he can operate virtually undetected for months. With a few basic precautions, most toll fraud can be eliminated.

Toll abuse, however, is another matter. No one knows the extent to which insiders use company toll for unauthorized purposes, but the amount is enormous. PBX and ACDs offer several features, discussed in this appendix, that help control abuse. Station restrictions, calls accounting systems, limits on features such as off-system forwarding, and, most of al, a firm policy can all reduce abuse.

All employees need to be alerted to the signs of toll fraud: multiple callers that hang up as soon as the phone is answered, requests to be transferred to an outside line, unexplained messages in voice mail, and unexpected all-trunks-busy conditions are signs that someone may be trying to hack your system, or may already have succeeded.

This appendix offers suggestions which, if properly applied, can reduce most toll fraud. Accompanied by a firm policy regarding personal use of long distance, much insider toll abuse can also be eliminated.

Block vulnerable area and country codes

The majority of toll fraud occurs to a narrow range of country codes plus the Caribbean. codes. If possible, block all international calling. If this isn't practical, consider establishing a narrowly-defined list of stations that have international access privileges, and restrict all other stations. If most stations must be able to place international calls, obtain a list from the IXC of the most vulnerable codes, and block these.

Restrict voice mail ports

The most common way for toll thieves to gain access to outside trunks is by dialing through voice mail. If possible, restrict voice mail from all trunk calls. If this isn't possible (for example because of out dialing to pagers) restrict outdialing to local calls if possible, if not pick the narrowest possible span.

Secure maintenance terminals

Be certain the passwords are changed from the default on both the switch and voice mail to a non-trivial password. Consider dial-back or other security devices for maximum insurance. Assigned the dial-up ports to a telephone number outside the DID numbering range.

Restrict voice mail transfer capabilities

Most voice mail systems have a code to enable callers to transfer to an extension numbers. Be certain that callers cannot transfer to an outgoing trunk, a tie trunk, or a trunk access code.

Secure trunk access codes

Trunk access codes allow callers to bypass the ARS and restrictions. Restrict the ability to dial trunk access codes the to a narrow range of stations, preferably only the switch room telephone. Make the trunk access codes at least six digits if possible, and be sure the number is not easy to guess.

Restrict off-system forwarding

Off-system forwarding should be prohibited if possible because it enables insiders to forwad a telephone to a long distance telephone number or to the operator dial the DID number from off-site, end complete the call. Block the feature if possible. If not, restrict it to a nearly-defined class of service.

Ensure physical security

Make certain all equipment rooms and telecommunications closets are locked and that access is limited to authorized persons. Be certain that access is restricted to the LEC's demarcation point.

Force voice mail passwords

To prevent hackers from capturing voice mail boxes, require a password length of at least six digits. If the system supports forcing periodic password changes, require users to change them quarterly.

Deactivate unused voice mailboxes

Hackers often attempt to capture unused voice mail boxes to communicate with others at no cost using your 800 lines. These attempts may be defeated by deactivating unused voice mail boxes and requiring the use of effective passwords on others. Effective passwords are at least six digits long and difficult to guess.

Review manufacturer's security manual

Most switch and voice mail manufacturers have produced a manual discussing security provisions. Review the manual and make sure the switch vendor has complied with all its recommendations.

Guard credit card numbers

Caution employees to shield their credit card numbers when dialing them into phones in public locations.

Shred old telephone bills

Many toll hackers operate by digging through the garbage to get lists of telephone numbers and credit card numbers. To defeat this, be certain that all telephone bills are properly disposed of after they have been paid.

Alert employees to signs of hacking

Instruct voice mail users to report unexplained messages or disappearance of messages that may indicate hacking attempts

Block trunk access codes

Be certain that trunk access codes cannot be dialed through voice mail. If possible, disable trunk access codes from all but a narrowly defined class of service.

Disable trunk-to-trunk transfer

If trunk-to-trunk transfer is not required, deactivate the feature. If not possible, restrict the feature to a narrowly assigned class of service.

Appendix C

Acronyms Used in This Book

AC	Alternating current
ACD	Automatic call distributor
AMIS	Audio messaging interface standard
ANI	Automatic number identification
API	Application programming interface
ARS	Automatic route selection
ATM	Asynchronous transfer mode
BBS	Bulletin board system
B8ZS	Eight-bit with zero suppression
BRI	Basic rate interface
CAS	Centralized attendant service
CDR	Call distribution recorder
CHAP	Challenge handshake authentication protocol
CLID	Calling line identification
CIR	Committed information rate
CO	Central office
CSU	Channel service unit
CTI	Computer telephony integration

DC Direct current

DHCP Dynamic host configuration protocol

DID Direct inward dial

DNIS Dialed number identification service

DSU Data service unit

DTMF Dual tone multifrequency

EIA Electronic industries association

EMT Expanded Metallic Tubing

ESF Extended superframe

FDDI Fiber distributed data interface

FEX Foreign exchange

FOD Fax on demand

HVAC Heating, ventilation, and air conditioning

IDF Intermediate distributing frame

IP Internet protocol

IPX Internetwork packet exchange

ISDN Integrated services digital network

IVR Interactive voice response

IXC Interexchange carrier

LAN Local area network

LATA Local access transport area

LEC Local exchange carrier

MAC Media access control

MDF Main distributing frame

MLPPP Multi-link point-to-point protocol

NIC Network interface card

NMS Network management system

NOS Network operating system

PAP Password authentication protocol

PBX Private branch exchange

PC Personal computer

PERT Program evaluation and review technique

PFT Power fail transfer

PIC Primary interexchange carrier

PPP Point-to-point protocol

PRI Primary rate interface

PSTN Public switched telephone network

PVC Polyvinyl chloride

QA Quality assurance

RAS Remote access server

RFP Request for proposals

RFQ Request for information

RMON Remote monitoring

SMDR Station message detail recording

SNMP Simple network management protocol

TAPI Telephony application programming interface

TCP/IP Transmission control protocol/internet protocol

TIA Telecommunications industry association

TSAPI Telephony service application programming interface

UPS Uninterruptible power supply

WAN Wide area network